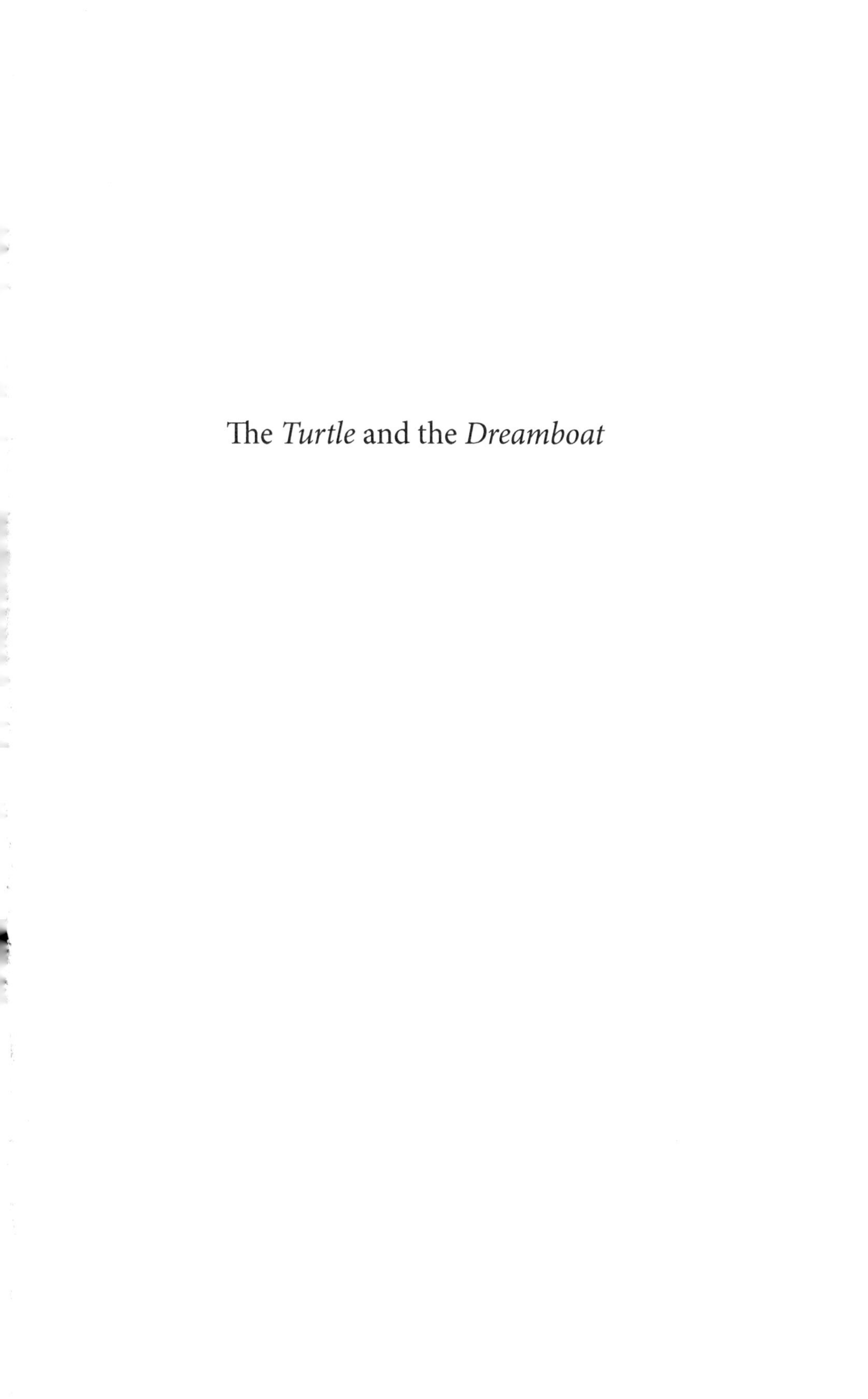

The *Turtle* and the *Dreamboat*

THE *TURTLE* AND THE *DREAMBOAT*

The Cold War Flights That Forever Changed *the* Course *of* Global Aviation

JIM LEEKE

Potomac Books
An imprint of the University of Nebraska Press

 Potomac Books is an imprint of the University of Nebraska Press.
Manufactured in the United States of America.

Library of Congress Control Number: 2021050540

Set in Arno Pro by Laura Buis.

For the airdales and the spooks

The record-shattering flights of two United States aircraft—the Truculent Turtle and the Pacusan Dreamboat—serve again to illustrate this shrinking world from the potential commercial or—more frightening—military viewpoints.

—*Cairns Post,* Australia

CONTENTS

ILLUSTRATIONS

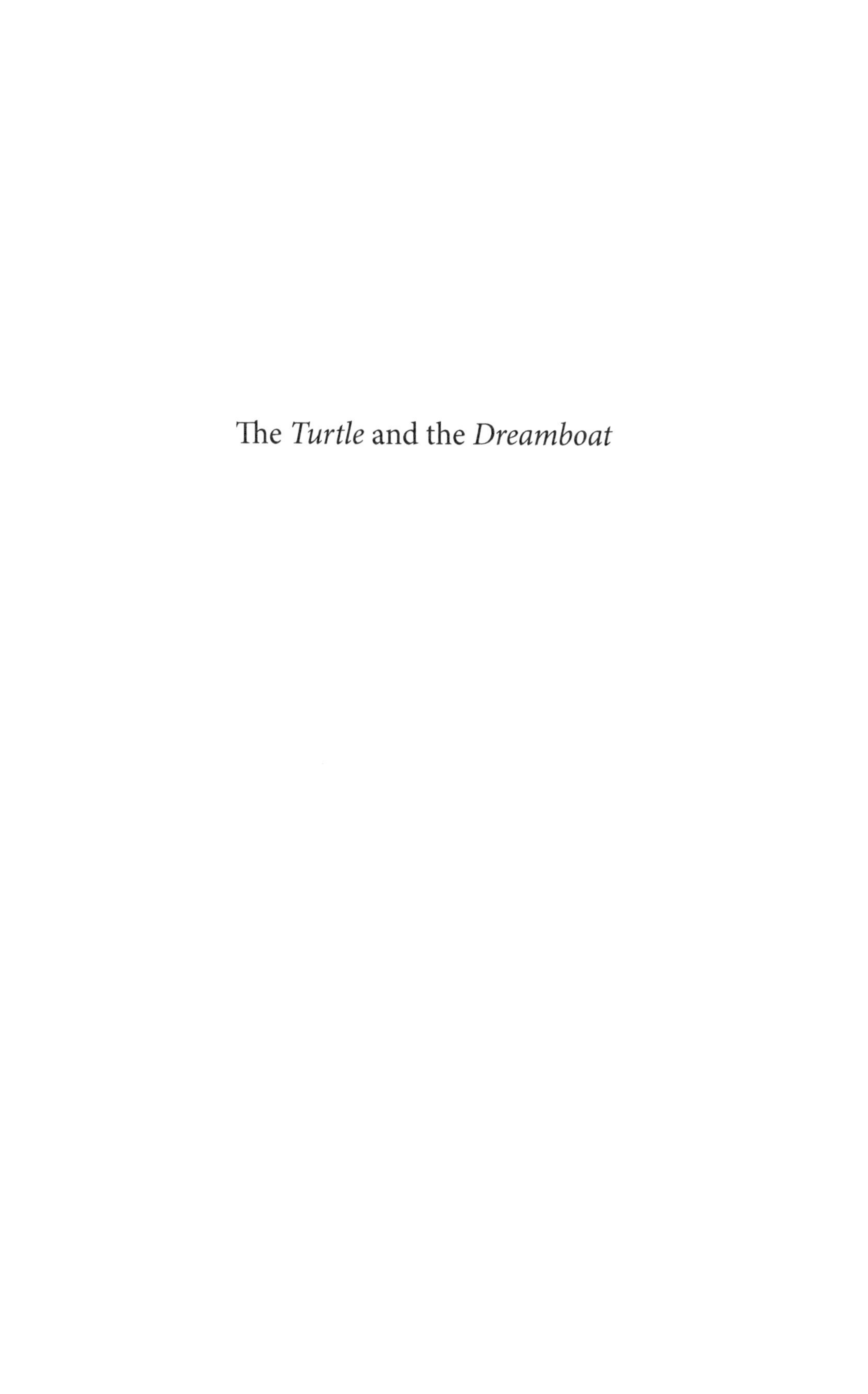

The *Turtle* and the *Dreamboat*

ONE

The Great Circle

WAVES OF BOMBERS FILLED TOKYO'S NIGHTTIME SKIES LIKE bats from a cave until two gleaming B-29s dropped new super weapons to end the war. As autumn approached, Americans in their capital heard the homecoming roar. Three of the bombers reached National Airport in mid-September 1945 with a general piloting each plane.

The airmen had set out to fly from Mizutani Airport on the Japanese island of Hokkaido nonstop to Washington DC, the first such flight ever attempted. Strong headwinds over Canada forced them instead to refuel at Chicago. The nearly six thousand miles they'd flown from the Far East to the Midwest nonetheless were an American, if not a world, nonstop record. Gen. Henry H. "Hap" Arnold, the five-star chief of the U.S. Army Air Forces (USAAF, commonly shortened to AAF), told journalists later in Washington that heavy bombers were so badly needed in the Pacific during the war that the four-engine Boeing B-29 Superfortress had gone straight from the drawing board into production. The long flight home, he said, was a service test the B-29 had never undergone.

Lt. Gen. Barney M. Giles commanded the mission. "But this was not a stunt and we weren't taking chances," he said. "The planes are too expensive. . . . We'll do it again." Maj. Gen. Curtis "Old Iron

Pants" LeMay, another of the arriving pilots, added that B-29 bombing raids had finished Japan. "The war would have been over in two weeks," he said, "without the Russians coming in and without the atomic bomb."[1]

A second trio of B-29s flew into Washington a month later from the Pacific island of Guam. The previous summer Superforts had lifted off from Guam and nearby Tinian and Saipan to drop tons of incendiary and high-explosive bombs on Japan. *Washington Star* correspondent W. H. Shippen Jr. described these raids as "the world's longest bombing run, all of 3,300 miles and more from here to Tokyo and back."[2] It was the distance from Seattle to the eastern tip of Cuba, nearly all over open ocean. The three returning B-29s reached Washington by the long way around, east to west, more than thirteen thousand miles. The chief of the Twentieth Air Force led the flight, which with refueling stops in India and Germany took sixty hours. The final leg from Frankfurt was the second nonstop, fixed-winged flight ever from Germany to America and the first since 1938. The War Department said the purpose of such flights was to "demonstrate the aeronautical smallness of the world and what can be accomplished with today's conventional bombers."[3] The three Superforts had forty AAF servicemen on board, each eligible for an honorable discharge.

Another flight of four B-29s reached the capital the first day of November. These planes flew sixty-five hundred miles nonstop from Hokkaido in twenty-seven and a half hours. All had flown missions against Japan during the war and had dropped supplies to starving American prisoners afterward. Coming home, they battled horrible weather and an eleven-hundred-foot ceiling over the Pacific. "We didn't know whether we would make it until we reached Kodiak, Alaska," said the general who led the flight. His pilots abandoned hopes of continuing past Washington to Puerto Rico only because of low fuel. "I wanted to prove a bomber could reach Washington from Japan and fly beyond it," said a disappointed colonel.[4]

The War Department made no official announcement, but the

B-29 hops to Washington in flights of three and four were now over. The next bomber would come alone, like a steely eyed marshal in a western movie.

THE SINGLE SUPERFORTRESS WAS UNPAINTED LIKE ALL the others. "Combat commanders recommended taking off the paint because fighters will become fifteen to twenty pounds lighter, and heavy bombers will lose seventy to eighty pounds," the *New York Times* explained during the war.[5] The lone bomber was called the *Dreamboat*, but the nickname wasn't intended to be romantic. "Dreamboat" was the call sign for navigational B-29s that had escorted fighter planes to and from Japan during the air campaign because, as Boeing's chief test pilot said, the big bomber "flies like a dream."[6] American fliers soon had applied the name to every B-29.

Col. Clarence S. Irvine of St. Paul, Nebraska, commanded this particular plane, which had flown about a dozen combat missions during the war. He later said it was called *Dreamboat* "because we were working on 44,000 pound bombloads when the war ended." Universally known in the AAF as "Bill," Colonel Irvine had entered the army as a private during the last days of the First World War and amassed a stellar record during the twenty-seven years since. He chewed rather than smoked a dozen to fourteen cigars a day. "During long flights," he explained, "smoking is out of the question because high test gasoline is being transferred to the flight tanks."[7]

Irvine physically resembled the little-known character actor Paul Ford, who a decade later would star in the new medium of television playing bumbling Colonel Hall on *The Phil Silvers Show*. The likeness went no further. Not even M.Sgt. Ernest Bilko would have dared to launch a scheme around this colonel, although Irvine himself had shredded miles of red tape and violated plenty of rules in supporting General LeMay's ferocious bombing campaign in the Pacific. With his chesty physique and crushed service cap, Irvine seemed more like a character from Milton Caniff's wartime *Terry and the Pirates* comic strip. Enlisted personnel who served under him later dubbed the Nebraskan "Turbine Irvine" for his energetic, hard-charging ways.[8]

The colonel began mulling a Guam-to-Washington flight in July 1945 while still fighting the Japanese. Irvine asked permission to make the attempt once the enemy surrendered, but the army stalled, "because a good many technical men said it couldn't be done," he later wrote. "On paper, by old-fashioned methods, it couldn't." Irvine's idea was to strip a B-29 of all nonessential gear for his proposed flight. "We wanted to demonstrate, in convincing fashion, that the B-29 still had plenty of room for exploitation of its powers—even when the dope sheets showed the flight couldn't succeed."[9] AAF commanders in Washington wouldn't believe a bomber could carry enough gas for the flight, so Irvine ordered his administrative assistant, Capt. Ruth A. Saltzman of the Women's Army Corps (WAC), to the Pentagon to argue otherwise. Armed with a briefcase stuffed full with technical charts and diagrams, she needed only an hour to sell the brass hats on Irvine's proposal. "Come when ready," General Arnold radioed.[10]

Yank aircrews had learned in 1945 how to strip nonessential weight out of a seventy-five-thousand-pound B-29 and give it the range to reach a target in distant Japan with a full bombload and get back home safely again. What Irvine and his men began calling the Dreamboat Project went further. Mechanics removed the auxiliary power plant, tail skid, and anything else the plane didn't need to stay aloft. A newspaper later described the *Dreamboat* as "a standard B-29 minus armor and radar," which was a simplification but not inaccurate.[11] When ready, it weighed sixty-six thousand pounds empty, but it needed over eleven thousand gallons of gasoline to reach Washington. With thirteen auxiliary fuel tanks topped off, the bomber weighed 151,500 pounds at takeoff.

Guam is the southernmost island in the Mariana chain, situated east of the Philippines and south of Japan. Japanese troops had captured Guam within days at the start of the war, but American forces recaptured it in 1944 and constructed three B-29 airstrips on the island. The *Dreamboat* was taking off for Washington from the strip called Northwest Field. Irvine planned to use all of an 8,500-foot runway of hard-packed crushed coral plus a 780-foot taxiway as well. The weather was good as he maneuvered into position at 5:30 on Monday afternoon, November 19 (2:30 a.m., Washington time). The pilots applied full power, released the brakes, and thundered ahead.

Northwest Field lay at Ritidian Point on the island's northern tip. Beyond the runway, the ground dropped dizzyingly away from a limestone cliff to the beach four hundred feet below. The bomber lifted off at 145 miles per hour into a fifteen-knot headwind. Asked later how much runway was left when his wheels left the ground, Irvine smiled and said, "We came out about even."[12] The *Dreamboat* dipped as expected below the plateau, picked up speed, and leveled off sixty feet above the sea. The plane then began a slow climb to a thousand feet and stayed there for four hours, until Irvine could begin easing up to ten thousand feet without wasting fuel.

THE SHORTEST DISTANCE BETWEEN TWO POINTS IS A straight line. Navigators call that line a great circle route, but in their case, it's actually an arc across the face of the earth; it only appears to be a line when viewed face on across a globe. An Australian newspaper later explained a great circle route's value: "Ships use flat maps—Mercator's projection—for navigation when steaming over small areas. The great circle effects little saving in distance over a small range. Where great distance is involved, the saving is substantial. . . . Aeroplanes pass over water and land, and the circle course is of great advantage."[13]

Unless it's east or west along a constant latitude, the relative direction of a lengthy great circle route continually shifts because of the earth's curvature. A plane flying north over the Arctic, for example, eventually heads south once past the North Pole. On a two-dimensional map, the great circle route from Guam to the District of Columbia starts northeast past Japan, bends due east below the Aleutian Islands, and then turns southeast down across North America. An army official quoted in many newspapers said the *Dreamboat*'s path to Washington was longer than those of the earlier flights from Hokkaido "because the B-29 did not take the short way home, over the top of the world."[14] The remark was inaccurate—the Arctic wasn't on the quickest path from the western Pacific to the nation's capital, as anyone with a globe could have determined—but it did eerily foreshadow coming events.

The Guam-to-Washington flight became a grueling series of physical and mechanical challenges for its special double crew. Irvine and his copilot, Lt. Col. George R. Stanley, started out stripped to their shorts and sweating. Hours later near Alaska, everyone wore cumbersome cold-weather gear. Having the crew work and sleep in shifts proved impossible because of the huge temperature swings. "First, it was so hot it was unbearable, then it was so cold we could hardly take that," Irvine remembered.[15] The crew battled through seven weather fronts. An electrical amplifier shorted out and disabled the auto pilot. With clouds above and below, a pair of navigators kept the *Dreamboat* on course by using what fliers called dead reckoning—calculating a position based on a fix taken earlier—and occasionally by radio direction finding. Every officer and noncommissioned officer (noncom) aboard had to operate at the top of his game despite the heat, the cold, and the lack of sleep.

A true great circle route would have taken the plane over Canada, but the B-29 approached North America farther south. The bomber encountered icing conditions five hundred miles northwest of Seattle. The Superfort carried no heavy deicing equipment and ordinarily would have flown over or under a freezing layer. "But on this deal, we figured it would cost us a hundred gallons or so to climb or to lose altitude and climb again and we didn't have the gas to spend," Irvine said.[16] He maintained his altitude until the plane broke through and wrote afterward that sometimes the snow was so sharp, it knocked the ice off the wings.

Irvine stopped the *Dreamboat*'s port outboard engine and feathered the four-bladed, sixteen-and-a-half-foot propeller shortly before sighting Cape Flattery in far northwestern Washington State, the first landfall since Guam. He had been running all four engines at 30 percent power to conserve fuel, but in the colder temperatures at higher altitudes, they were beginning to vibrate so violently it seemed they might "shake the plane to pieces," he said.[17] "The fourth engine, therefore, was stopped and the power on the other three increased to warm them," Irvine added. "The result was to keep the plane flying with the same fuel economy but without the dangerous vibration from cold engines."[18] The *Dreamboat* continued on three engines most of the way across the continent.

The time in the United States was early Tuesday, November 20, the bomber having crossed the international date line in the Pacific. The unprecedented flight continued eastward past Spokane, Washington; over the Rocky Mountains at 12,500 feet in the predawn darkness—"very unpleasant because we were too close to the top and in thick soup all the time," Irvine remembered—then down to 8,000 feet; across the northern states; past Great Falls, Montana, and Bismarck, North Dakota; and on toward the Great Lakes.[19] "Unlike other B-29 flights from the Pacific, this one was unannounced by the War Department until the bomber passed over La Crosse this morning," the *Washington Star* reported. "The picked crews received special training and equipment, and air-sea rescue stations were alerted along the route to insure safety of operation."[20]

The army figured that by the time it passed the Wisconsin city at 8:30 a.m. central standard time, the *Dreamboat* had beaten the old distance record set by British fliers in 1938. The government radio and weather station in the area hadn't been informed, and no one on the ground apparently noticed the flight, but the *La Crosse Tribune* found cause to celebrate. "Well, that puts us in the record books of aviation for keeps," its editorial declared. "Can't you just fancy that they'll tell it, over and over again, how the B-29 broke a record at La Crosse, or over La Crosse, forgetting that the bloomin' plane kept on going to Washington?"[21]

The bomber passed above western Pennsylvania three hours later with only the Alleghenies standing between it and Washington. "We reached Pittsburgh before we were sure that we could finally make Washington," Irvine recalled. "The last 58 minutes seemed like a year because tension and lack of sleep affected all of us."[22] At last, with the engine switched back on and the feathered prop again spinning, the colonel guided the *Dreamboat* down toward National Airport. Irvine kept the bomber's nose nearly level for landing rather than pitching it slightly upward in the normal configuration. The shallow angle kept the remaining gasoline forward in the fuel tanks and avoided the possibility of starving the engine in the final critical moments. Irvine set the plane onto the runway "gently as thistle-down" at 1:35 p.m. eastern standard time.[23]

The B-29's aluminum skin flashed in the chilly sunshine as he taxied toward the Air Transport Command (ATC) Terminal. The *Dreamboat* had exceeded the recent Hokkaido flight by over sixteen hundred miles, winging home in thirty-five hours and five minutes. "As soon as the great propellers stopped swirling, bomb bay doors opened and crew members scurried down ladders," an International News Service correspondent wrote.[24]

The flight across one-third of the globe set a nonstop distance record. The plane still had three hundred gallons of gas in its tanks and could have continued flying another sixty or ninety minutes. Irvine popped out of the plane first, wearing a flight suit and crumbled service cap, the unlit stub of a cigar clenched in his teeth. Nine other equally rumpled airmen tumbled out behind him. A group of officers and government officials watched as General Arnold grinned and pumped the pilot's hand. "Hullo, Colonel," he said, "you've come a long way."[25] Irvine laughed and chewed his stogie, appearing "in excellent spirits despite a two-day growth of beard."[26]

Dreamboat actually had flown somewhat farther than the distance that went into the official record books. The army said the flight had covered 8,198 miles, but the Fédération Aéronautique Internationale in Paris certified a straight-line flight of 7,916 miles

(12,739 kilometers). This was the distance of the great circle route between the airfields at Guam and Washington without drift, deviation, or correction, which was how the federation calculated such things. "That was a grand flight, beating the world record by over a thousand miles," Arnold told Irvine on the tarmac. "Did you have any trouble?" The colonel replied as if the *Dreamboat* had hopped in from Atlanta or Buffalo: "No trouble at all."[27]

Irvine's bleary-eyed crewmen lined up for the cameras beside their gleaming B-29. General Arnold had learned to fly in 1911 at the Wright brothers' school in Dayton, Ohio, as one of the army's two first pilots. Now he pinned Distinguished Flying Crosses on the American airmen who had flown in nonstop all the way from the Marianas. Irvine's award was his second. "You and your crew did a grand job on a grand flight," the general told him. He added with a grin, "I have been talking about a 10,000-mile airplane. It seems like it is almost here."[28]

The remark was an understatement. "Actually, it is not only General Arnold's 10,000-mile plane that is 'almost here,'" the *Washington Star* editorialized. "'Almost here,' too, in the sense that they can be designed right now, are 'space ships' capable of traveling at incredible altitudes above the earth's atmosphere and pilotless craft of supersonic speed."[29]

General Arnold knew that Irvine and his *Dreamboat* crew were only getting started. He likely knew, too, that they and the AAF weren't alone in their ambitions. With a postwar unification of the U.S. armed forces in the works, the U.S. Navy wasn't going to stand politely aside and applaud the army's accomplishments. Indeed, a band of naval aviators right there in Washington soon would start planning a record-smashing long-distance flight of its own. Over the coming year, an emerging interservice aerial rivalry would span continents, traverse the Arctic, create a new global route for commercial aviation, and prove, one way or the other, that Hap Arnold was more prescient than he realized.

TWO

Unification

HARRY S. TRUMAN ASCENDED TO THE PRESIDENCY IN APRIL 1945 following the death of Franklin D. Roosevelt. The former vice president then ended the war in August by ordering the use of the atomic bomb, a weapon he hadn't known existed until becoming commander in chief. Now in late autumn, he was dealing with the difficult issues of peace.

The *Dreamboat* landed at National Airport the same day Truman announced several changes in the top echelon of America's armed forces during a regular weekly news conference at the White House. Two changes were especially noteworthy: General of the Army and Chief of Staff George C. Marshall would soon retire and be replaced by General of the Army Dwight D. "Ike" Eisenhower, the commander of American forces in Europe. Chief of Naval Operations (CNO) Fleet Adm. Ernest J. King would retire, too, with Fleet Adm. Chester W. Nimitz replacing him. Both departing officers were past normal retirement age, each having remained on duty during the war through an act of Congress.

"The President said both Gen. Marshall and Admiral King had requested retirement after Japan surrendered but that he had prevailed on them to stay until now," the Associated Press (AP) reported.

Hap Arnold wanted to be relieved, too, but the president hadn't yet agreed. Truman went out of his way to praise Marshall, calling him "the greatest military man that this country ever produced—or any other country, for that matter."[1] Within weeks, he would nominate Marshall as secretary of state and see him unanimously confirmed by the Senate.

As the afternoon press conference was ending, a reporter asked Truman about what the *New York Times* called the "highly controversial Army-Navy merger question."[2] The issue had simmered for decades. Congress studied military reorganization no fewer than twenty-six times between the world wars without acting until World War II finally impelled a push for real change. Members of a Senate subcommittee interviewed eighty top military officers in four military theaters in late 1944. According to the *Washington Star*, their report—written in April 1945 and initially classified—recommended establishing "the Army, Navy and Air Force with equal status under a Commander of Armed Forces who is responsible to a Secretary of Armed Forces."[3] The senators backed an overhaul of the nation's longtime military structure, in other words, under which the War Department currently controlled the army while the Navy Department ran the sea service.

According to the subcommittee's report, the two coequal departments were organized "along cumbersome and inefficient lines which hindered rather than facilitated cooperation." The once-secret document added: "At the end of 3 years of war the special committee has observed that even in areas where unity of command has been established, complete integration of effort has not yet been achieved because we are still struggling with inconsistencies, lack of understanding, jealousies and duplications which exist in all theaters of operations. That these handicaps have been overcome in any degree is due to the stature of our leaders at home and abroad."[4]

The damning report further declared that legislation to create a single department to oversee all the armed forces should be enacted within six months of Victory over Japan Day—that is, by February

1946. Although not binding, the recommendation sparked strong and sometimes heated debate. "The great majority of the Army officers and almost exactly half of the Navy officers, whose views were heard, favored the single department," the report said, "although much difference of opinion was expressed as to the details of the form which it should take."[5]

Truman believed in organizational integration of the military services. (He would address racial integration with an executive order desegregating the armed forces in 1948. During and shortly after the war, however, African American army pilots still flew in segregated squadrons, while the navy, shamefully, had none at all.) The Missourian had advocated service integration while a senator and vice presidential candidate in 1944.

"Proof that a divine Providence watches over the United States," he had written in *Collier's Magazine*, "is furnished by the fact that we have managed to escape disaster even though our scrambled professional military set-up has been an open invitation to catastrophe." The obvious first step, Truman added, was consolidating the army and navy "under one tent and one authoritative, responsible command."[6] Now fifteen months later, a White House reporter was asking about a merger. The assembled journalists didn't quote the new president directly, but Truman "indicated that he would soon send a message to Congress recommending unification of the land, sea and air forces forthwith."[7]

A merger or unification—newspapers used both words interchangeably—was particularly important to the U.S. Army Air Forces. "If the USAAF remained subordinate to the Army, its wartime record and the atomic bomb guaranteed that its status would change," a military historian wrote half a century later. "The atomic bomb had altered the nature of warfare."[8] Many officials in Washington shared Truman's view on unification. Secretary of War Henry Stimson recommended it in 1944 but only once the war ended. The conflict made it abundantly clear, Stimson said, "that voluntary cooperation, no matter how successful, cannot under any conditions of warfare, and particularly under triphibious warfare, be as effective in the handling of great military problems as some form of combination and concentrated authority at the level of staff planning, supervision and control."[9]

Ike Eisenhower likewise shared Truman's belief in integration. The week before the *Dreamboat* reached Washington, the five-star general told the Senate Military Affairs Committee that America's postwar security would become a "patch-work improvisation" if the government didn't place the army, navy, and AAF under unified command. "With integration we can buy more security for less money," he said. "Without it we will spend more money and obtain less security."[10] Eisenhower made the point again to an American Legion convention in Chicago a few hours after Truman's remarks at the White House. "A strong America is a trained and an integrated America," he told the Legionnaires. "Nowhere is that integration more necessary than in our armed forces. We must not think, primarily, in terms of ground forces, naval forces, air forces. We must think in terms of coordinated action."[11]

Gen. Joseph Stilwell and much of the army agreed. "Vinegar Joe" commanded the China-Burma-India (CBI) campaign in 1942–44; then he led the Tenth Army on Okinawa. In early December 1945, Stillwell compared military services to fire departments. "The fire fighters are organized in chemical companies, hose companies, ladder companies, and in some cities, fire boats are used," he said during a radio debate. "But when the alarm rings they all rush together to the one fire. And it's the fire chief who decides what piece or pieces of equipment will be used."[12] Bill Irvine employed a similar analogy while backing the army line later in the month during an event in Philadelphia. "Colonel Irvine agreed that unification of the armed services on an equal basis is this country's best defense," the *Philadelphia Inquirer* reported. "He emphasized that B-29s should be placed strategically on a world 'beat' as 'scout cars of a police force.'"[13]

While acknowledging that structural changes such as interservice training might be necessary, the navy largely and often vehemently rejected any merger. Admiral Nimitz opposed it during a radio broadcast; Adm. Henry Hewitt, commander of U.S. naval forces in Europe, and other officers echoed their CNO. "The retention of a strong Navy is of the utmost importance to our national security and to the playing of our part in the United Nations Organization," Hewitt later told the Senate Military Affairs Committee. "This Navy should not be placed under the control of those who do not understand its functions and problems."[14]

Military historian David Alan Rosenberg sums up the fierce interservice conflict this way: The army and AAF "argued that service missions should be defined in terms of the medium in which each service operated, i.e., land, sea, and air, and that duplication in weapons systems should be eliminated. . . . The Navy responded that function, not weapon systems, should determine the role and composition of each service."[15] Nimitz and his admirals weren't about to turn over any control voluntarily to the AAF or the independent air force they saw on the horizon. As a later editorial in an Iowa newspaper rightly observed, "Neither service wants the other to become predominant in the public mind as the progressive, air-minded arm of the national defense."[16]

President Truman had fought in France as a field artillery officer during World War I and wasn't cowed by gold braid and silver stars nearly thirty years later. He sent Congress a six-thousand-word message before Christmas, asking for legislation to combine the Army and Navy Departments under a new Department of National Defense. "He wanted to break up the power of the West Point and Annapolis cliques, to make the armed services more democratic—a noble aspiration, many around him agreed, but impossible, they felt," writes Truman biographer David McCullough. "It was his duty to send the message, he said, because it represented his conviction."[17]

Stubborn as a Missouri mule, Truman proposed that a civilian head the new department, with an undersecretary for each of the land, sea, and air forces. "The Navy should, of course, retain its own carrier, ship and water-based aviation which has proved so necessary for fleet operation," he wrote. Addressing a concern expressed by Admiral King and others, the president added, "There is no basis for the fear that such an organization would lodge too much power in a single individual—that the concentration of so much military power would lead to militarism."[18]

The navy disliked Truman's message despite his nod toward maintaining fleet independence, "fearing that a separate air force might supplant the fleet as the nation's first line of defense and 'swallow up' naval aviation—thereby placing its postwar maritime strategy in jeopardy," writes air force historian Warren Trest.[19] The navy nonetheless issued a gag order to naval and Marine Corps officers "to preclude their expressing opposition in public to the President's recommendation for a department of national defense," the *Washington Star* reported. "The muzzling order, sent out immediately after the release of the merger message, permits officers to express their views if called as witnesses before congressional committees."[20]

The controversy didn't die there but continued to rage two years later as Truman signed the National Security Act of 1947. The act formed the Department of Defense and created the U.S. Air Force, as historian Trest writes, "without uprooting the Navy's sovereignty over naval and Marine Corps aviation."[21] The conflict between the services rolled on despite the assurance. Infighting peaked but hardly ended in 1949 when newly appointed defense secretary Louis A. Johnson abruptly cancelled construction of the supercarrier USS *United States* a few days before the keel was to be laid without informing either CNO Adm. Louis E. Denfeld or navy secretary John L. Sullivan.

"Secretary Sullivan resigned in anger, and the so-called 'Revolt of the Admirals' soon rocked the nation's capital," a naval historian writes. "The aborted aircraft carrier became the crux of the entire inter-service issue of unification and postwar strategy."[22] The navy fought back during a congressional inquiry with accusations and insinuations about malfeasance in the air force's billion-dollar Convair B-36 Peacemaker bomber program. These allegations were unfounded. Gen. Omar Bradley, the highly respected chairman of the

Joint Chiefs of Staff, scorched the navy during his testimony with a pithy sports analogy: "This is no time for Fancy Dans who won't hit the line with all they have every play, unless they can call the signals."[23] The revolt collapsed, and Denfeld was relieved as CNO. "But inter-service rivalry remained bitter in the decade that followed, even if it stayed out of the public limelight," says a modern naval historian.[24]

All this nastiness still lay several years ahead in 1946. The inter-service battle meanwhile continued like an especially brutal army-navy football game, albeit one played on a much bigger field with far greater stakes.

THREE

Cross Country

THE AAF USED THE *DREAMBOAT* TO WAVE THE FLAG FOR army aviation following the flight from Guam. Irvine and his crew toured the country in the B-29 on a postwar bond campaign, with the colonel receiving a hero's welcome December 5 at his Nebraska hometown. Longtime friends in St. Paul still used his old nickname "Mutt." The little community's only other celebrity was former star pitcher Grover Cleveland Alexander, a member of baseball's Hall of Fame. "Alex the Great" came home only occasionally, his fame tarnished by a long struggle with alcohol. But townsmen had no reservations in greeting Irvine, a true favorite son whose visits and exploits always received coverage in the area's two weekly newspapers.

Irvine stayed in the army as a mechanic following World War I and rose quickly to sergeant first class at Kelly Field, Texas. "The aviation branch is the best of the army to his way of thinking, and if a man must be in the army that is what he desires for his work," the *St. Paul Phonograph* told readers in 1919.[1] Mutt qualified as a pilot the following year, earned a commission in 1921, and served several years in the reserves before returning to active service. As a member of the First Pursuit Group at Selfridge Field, Michigan, Irvine was one of six army pilots loaned to Hollywood to fly in the 1927 motion picture *Wings*, in which he portrayed a German flier called Kellerman. "Few persons who view the thrilling air spectacle, 'Wings,' which opens at the Capitol theater next week . . . may be aware that one of the outstanding aviators in the film, depicting the role of a German war ace, is a young man who was born and raised in this community," the *Phonograph* proudly noted.[2]

Irvine made news again in 1929 by crash-landing his pursuit plane twenty feet off a beach in the Philippines after his motor quit in a cloud of black smoke. "His action in refusing to risk the lives of bystanders, although he might have done so without killing anybody and without risking his own neck, has brought him much favorable comment," the *Manila Daily Bulletin* applauded.[3] Back in the United States during the 1930s, Irvine let home folks know whenever he was nearby by buzzing his old hometown to say hello. "It's a free air circus that no one wants to miss," the *Phonograph* once observed.[4] "Once they dismissed the whole grade school and we stood on the grounds while Capt. Irvine put on a daring show," a former student remembered. "As he would barely brush the tree tops, we all would scream with delight. One time he hit the brick on the chimney and we thought that was great!"[5]

The popular flier crashed again in 1938 during a fierce summer storm in upstate New York but walked away with only a sprained wrist. A year later, he and another army pilot set records for altitude and speed on the same day, flying an early B-17 Flying Fortress. During World War II, the weekly papers printed article after article describing his whereabouts, accomplishments, and medals. While stationed in Texas, Irvine had secretly wed a widow with a young son in 1926, but they seemed rarely to live together and had divorced during the war. The locals never tired of reading about Irvine and welcomed him home now after his epic flight in the *Dreamboat*.

"A delegation from St. Paul and the members of Col. Irvine's family met the 'Dreamboat' at the Grand Island [Nebraska] Army Air Base when it landed at 10:21 that morning," the *Phonograph* reported. "Col. Irvine and his party were escorted to St. Paul in a motor caravan for a gigantic parade."[6] An army band led five hundred soldiers in a procession that also included local veterans, the high school band, and every schoolkid in town. Radio station WOW in Omaha broadcast part of the proceedings. The crowd consumed more than a ton of beef at a noon barbeque, and a dance followed that evening.

Irvine didn't linger amid the adulation. He was in Seattle the following Monday, having flown the *Dreamboat* there from Bridgeport, Connecticut, in a record ten hours. The colonel gave journalists in the Emerald City a grim account of what a Third World War might look like, a narrative the AAF wanted Americans to hear. "Using 30 B-29s equipped with what he called 'the latest type of bomb' Col. Irving declared he could have wiped out every industrial center in the United States in a single sweep."[7] Irvine took the *Dreamboat* to Southern California the following day, December 11, for a flight to win fresh headlines. He reached Burbank after a record run of over a thousand miles in fewer than three hours but remained on the ground less than four hours before roaring aloft again, bound for New York. Later, as the big bomber passed over his home state, Irvine radioed down to the Grand Island airfield: "This is St. Paul's Irvine taking another crack at a record."[8]

The *Dreamboat* averaged 450 miles per hour, pushed along by strong tailwinds, and topped 550 in flying nearly twenty-five hundred miles to Brooklyn. The average was 18 miles per hour faster than a mark set three days earlier by an army XB-42 Mixmaster, an experimental bomber with propellers mounted in the tail, that flew a shorter route from Long Beach to Washington. The *Dreamboat* also broke a mark set in May by a P-51 Mustang fighter flying from Inglewood, California, to LaGuardia Field in Queens. Irvine flew through a patch of rough weather while descending toward Brooklyn, buzzed Floyd Bennett Field and dipped his wings at three thousand feet to establish his latest mark, and then turned north and landed at LaGuardia. A hundred or so people greeted the B-29 at the ATC hangar.

"There was a new cross-country flight record today—less than

5½ hours from Burbank to Brooklyn—set by the B-29 Superfortress Dreamboat just before midnight last night," the *Brooklyn Eagle* reported. "It looked like a record which, very likely, would not stand long."[9] The *New York Daily News* added that Irvine and his crew "took their achievement calmly. It was their third record flight in two days, and came on the heels of their epochal 8,193 miles nonstop record distance flight from Guam to Washington last month."[10]

The colonel told assembled newshounds that his plane had "no special features" and added that any B-29 might have set the same record.[11] Irvine's ten-man crew, his WAC assistant, and three civilian passengers all wore oxygen masks and heavy flight suits throughout the flight. The bomber flew most of its run at thirty-five thousand feet and nearly a thousand feet higher when crossing the Rockies, the longest any plane had flown at that altitude. "My gosh, it was cold up there," said crew chief M.Sgt. Dock E. West of Tazewell, Tennessee.[12] Irvine agreed, calling the flight "clear, cold and rough as hell." He had another blunt answer when asked how much gas the *Dreamboat* had left after landing. "I don't know how much, but it was enough," he said. "Anything above five gallons is enough on a flight like this."[13] Newspapers reported that of the 6,200 gallons taken aboard, the *Dreamboat* had 450 gallons left at LaGuardia.

Irvine called himself "just a glorified chauffeur" and said little extra preparation had gone into the flight.[14] "All we did was change a few spark plugs and take some stuff out to make room for gas," he told journalists. "We had no desire to kid ourselves with souped-up equipment. This run was a pretty rough test of our equipment. We ran the engines at full power all the way."[15]

The gruff Nebraskan thought the record-setting run offered a glimpse into the future of transcontinental aviation. "A coast-to-coast flight in four hours can be achieved through work and money," he said. "It won't require much change in the planes we have now for such a flight."[16] Elsewhere the colonel predicted that the data he and his crew had gathered during the flight would be useful for designing high-altitude commercial airliners. He added that four-hour coast-to-coast flights were possible within a year. "Such schedules can be put into operation 'whenever we are ready to spend the money,' the colonel said. By such flying it would be possible to leave New York at noon and keep a 1 p.m. luncheon date in San Francisco the same day."[17]

Irvine penned an article for the International News Service in which he noted that his crew had experienced no problems with its complicated instruments or anything else. He also praised the anonymous aviation workers who had contributed to the *Dreamboat*'s latest triumph. "If it hadn't been for all the Joe Doaks working in the Hamilton Standard Propeller Works, the Boeing plant, the Bulova Watch factory, and a hundred other shops around the country, something would have conked out at some time," he wrote. "My hat, and the hats of my crew, are off to them!"[18]

But for once the colorful and always-quotable colonel wasn't the media darling after the *Dreamboat*'s Burbank-to-Brooklyn jaunt. The spotlight instead shone on his assistant, Capt. Ruth Saltzman of Washington DC, the only woman on the flight. Journalists naturally took an interest in the slender, immaculate captain accompanying the colonel.

Ruth Saltzman had enlisted in 1943, flown to Saipan with Irvine in October 1944, and served with distinction in the Marianas as the only woman in the Twentieth Air Force. She had been Irvine's trusted and highly capable assistant (the newspapers usually said secretary) ever since. The exact opposite of her stocky, untidy boss, the captain sometimes straightened the colonel's tie so that he looked as if he had at least completed basic training.

The *New York Daily News* interviewed Captain Saltzman at the Waldorf-Astoria Hotel the day after the flight. She explained that she worked in supply and maintenance out in the Pacific. "Hard work, yes, but wonderful." She thought the record-setting flight from Burbank had been terrific and described dining on a sandwich and milk in the bomber's pressurized cabin. "A girl in a bomber must always wear slacks," Saltzman said. "Up at 35,000 feet I also had to wear a fleece-lined coat and a parka." The unmarried captain deftly turned aside a question from the female reporter about proposals from all those eligible men out in the western Pacific. "Everyone was too busy, never got around to it," she said. "Social life was out. We had curfew at 10 P.M."[19]

Saltzman was one of three captains in her family. One brother served in the ATC, and another was a flight engineer on a B-29 in the CBI. The three battle stars on her own service ribbons, she said, were merely for working in combat areas. With the war over, she looked forward to sitting down and relaxing for once.

Colonel Irvine, meanwhile, was planning ahead. During December, he met with representatives of the Engineering and Procurement Divisions at Wright Field in Dayton to talk about converting a standard B-29—that is, the *Dreamboat*—into what he called a "climatic" aircraft with a range of ten thousand miles and capable of flying anywhere in the world.[20] Not long afterward, he wrote to Maj. Norman Pershing Hays, an AAF navigator who had served under him on Saipan. He asked Hays to begin planning a transpolar flight from Honolulu, Hawaii, to Cairo, Egypt. Mutt Irvine never thought small.

FOUR

Neptune

THE LOCKHEED AIRCRAFT CORPORATION ROLLED OUT A new navy patrol bomber ten days after Bill Irvine's Burbank-to-Brooklyn dash. Dubbed the P2V Neptune, it was the "first of a long series of the twin-engine long-range plane, designed for high speed over great distances."[1] While important to the navy's emerging postwar plans, the plane's origins lay in America's last uneasy days of peace.

The navy lacked any land-based patrol plane before the war, and redesigning army planes proved unsuitable. "On the eve of the United States entry into the war," writes an aviation historian, "it had become apparent that the US Navy would need a land-based patrol bomber with more range, greater armament load, higher level and climbing speeds and slower approach and landing speeds than provided by the Lockheed Hudson and Ventura."[2]

The twin-engine Hudson—"known affectionately as 'Old Faithful' throughout the RAF"—went primarily to British and Australian naval and air forces already fighting Nazi Germany.[3] American, British, and commonwealth forces later flew the Ventura, the Hudson's successor. The Ventura's designation was PV-1: *P* for patrol and *V* for Vega Airplane Company, the Lockheed subsidiary that built it. Vega then began preliminary work on the navy's as-yet-unnamed new patrol plane at Burbank. Vega's vice president of engineering was Mac Van Fleet Short, a World War I aviator and former airmail pilot and barnstormer. Short had a simple philosophy concerning airplane design: "Get out with the customer and find out what his problems are. Then try to help him with those problems."[4]

Short authorized internal design studies on the navy patrol plane on December 6, 1941. Japan catapulted America into the world conflict the next morning by bombing the U.S. fleet at Pearl Harbor. Vega soon shunted the fledgling project aside to meet the navy's immediate requirements but never entirely abandoned work on it. After the Battle of Midway halted the Japanese advance in the Pacific in June 1942, the navy asked Lockheed for patrol bombers with more range and lethality. Although the PV-2 Harpoon, an enhanced version of the Ventura, would take to the air in 1944, it was already apparent to everyone by 1943 "that what was needed for the Pacific war was an entirely new, much bigger and much more powerful land-based patrol plane."[5]

The navy signed a letter of intent in February 1943 to develop two experimental versions of the no-longer sidetracked plane. The company named Robert A. Bailey as the project engineer for what became known as the Neptune. "Every detail of the proposed new plane was studied on drawing boards, under Mr. Bailey's direction, in the months of hard work that followed," a newspaper recalled.[6] The young engineer showed the first blueprints to the navy in August.

Bailey was not yet thirty years old but already an experienced designer. Born in Livingstone, Montana, he began building gliders while a teenager and later soared to eighteen thousand feet without oxygen. He joined Lockheed in 1937 after earning a degree in aeronautical engineering from the Curtiss-Wright Technical Institute. Three years later, he designed a pilotless plane controlled by radio from the ground for towing targets during aerial gunnery practice. He next helped to equip a dozen B-17s with fourteen gun turrets apiece instead of the usual six, turning them into real flying fortresses. These heavily armed gunships protected more lightly armed

B-17s from enemy fighters during daylight raids over Germany. Bailey's career clearly was ascendant at Lockheed.

The P2V Neptune project really got moving once the navy issued contracts in April 1944 for two prototypes and fifteen production planes. Lockheed built them itself rather than Vega, which had been absorbed into the parent company. Lockheed president Robert Gross believed the new Neptune would endure, calling it "an airplane that has some stretch to it."[7] The result, a naval publication later declared, was "a twin-engine craft with an efficiency unattainable in four-engine designs. . . . Particularly, it has the low-speed characteristics which enable it to get in and out of small fields, yet it has a high cruising speed and a top speed well in excess of 300 m.p.h."[8] The first prototype, designated XP2V-1, rolled out of the Burbank factory in early 1945 and made its first test flight May 17. "It made me feel that all the long hours, the headaches, and effort put into the building of the plane had been more than worthwhile," Bailey recalled. "It is a great thrill on such an occasion."[9]

CDR. THOMAS D. DAVIES WAS ALMOST AS CLOSELY INVOLVED with the new Neptune as Bailey was. The patrol plane class desk officer at the Navy Department's Bureau of Aeronautics in Washington, Davies was also the Neptune's project officer. No naval aviator was better suited to the job.

Davies grew up in Cleveland with a connection to seafaring; his father developed gantry cranes used in shipbuilding. Tom attended the Case Institute of Technology for two years before entering the U.S. Naval Academy in 1933, three years behind his older brother. While working in the academy's physics lab, the brilliant young midshipman developed a stereoscopic range finder later used on big-gun warships during World War II and the Korean War. Classmates remembered Davies as a designer, an artist, and an academic savior for other "mids" who lacked his scholastic abilities. "T. D.'s disposition is one that allows him to obtain the most out of life, for little upsets and inconveniences never worry him."[10]

After receiving his commission in 1937, the same year Bailey joined Lockheed, Davies served on the cruisers USS *Portland* and USS *Wichita*. He then transferred into naval aviation shortly after America entered the war and finished his flight training in late 1942. The navy assigned him as the executive officer (second in command) of Bombing Squadron 129, designated VB-129, which flew Venturas. His new outfit shifted south to fly antisubmarine missions out of Brazil, protecting shipping from German U-boats lurking off South America.

The afternoon of July 30, 1943, as a lieutenant commander, Davies flew his Ventura in a sweep ahead of an Allied convoy. At ten minutes past two o'clock, the plane surprised *U-604* cruising on the surface a hundred miles off Maceío. The U-boat's captain opened fire. "Davies straddled him with four Mark-47 bombs spaced 75 feet apart," naval historian Samuel Eliot Morison wrote. "*U-604* submerged, went slowly ahead, then broached at a steep angle, screws fanning the air, and went down. Davies returned to base and reported a sure kill."[11]

Although not immediately sunk as Davies believed, the U-boat was badly damaged. Attacked again by other aircraft, the crew abandoned *U-604* after rendezvousing with two other boats farther out to sea. Davies received the Distinguished Flying Cross for his attack. While in Brazil, he also somehow found time to learn to read and write Portuguese. After completing his tour with VB-129, Davies stayed in the country to command a U.S. training unit, wrote an instruction book for Brazilian pilots, and translated maintenance manuals as well, all of which earned him the Brazilian Order of the Southern Cross. By early 1945, he was ordered to a new duty post in Washington.

THE NAVY PUBLICLY UNVEILED THE P2V IN BURBANK THE week before Christmas 1945, a month after the *Dreamboat*'s flight from Guam to Washington. Lockheed officials declared that their plane possessed "the greatest range, fastest speed and heaviest armament of any patrol and search bomber yet developed." They hailed the Neptune as "a new kind of airplane for the new era in world relationships—a peace patrol plane to guarantee law and order."[12] It was a handsome aircraft with pleasing lines and good proportions. *Popular Mechanics* later called it "aerodynamically one of the cleanest airplanes ever built."[13]

Unlike gleaming Superfortresses, the Neptune got a coat of paint so deep a blue it was sometimes mistaken for black. Powering the plane were two, twenty five-hundred-horsepower Wright R-3350 Cyclone 18 engines, the same type used on the B-29. "On the P2V a particularly compact and clean installation has been achieved, the exhaust outlet being well to the rear on the nacelle sides," an aviation magazine reported.[14] The Neptune had only two engines but wasn't a small plane—about the same size as a four-engine B-17 but far smaller than a B-29. Its wingspan was a hundred feet, its length slightly over seventy-five feet, and its high tail rose twenty-eight feet above the tarmac. The Neptune could carry a crew of seven, including gunners for top and tail turrets, plus an impressive array of armaments that included torpedoes, rockets, depth charges, or four tons of conventional bombs. And it could deliver atomic weapons, a capability the navy always took pains to point out.

The P2V held the prospect of incredible versatility. "Engineers say the plane can perform equally well on high altitude photographic or low level rescue missions, or can be transformed quickly into an attack bomber," the AP reported.[15] The plane had what was called a varicam tailplane, or horizontal stabilizer, which included a mechanism for varying the curvature of the surfaces to adjust trim; this in turn let a pilot balance a heavy load and keep the plane flying levelly. The Neptune topped three hundred miles per hour and climbed to a service ceiling of twenty-three thousand feet.

The United Press reported that the plane had "the carrying load and range of World War II four-engine bombers such as the B-17 and B-24." But the most impressive aspect was its range—five thousand miles when equipped with extra fuel tanks. "As an example of its range, from Guam it could fly over Wake Island, Tokyo, Manila or Rabaul, New Britain," the press added, citing locations familiar from the war. "Based on Manila, it could patrol Singapore, Shanghai or Hanoi, Indo-China."[16] Its multiple capabilities led to a widely syndicated headline declaring the Neptune the "Fastest, Farthest, Fighting'est" plane in the fleet.[17]

But the war was over, and thousands of surplus aircraft were bound for the boneyards. Despite the successes of its fast carrier battle groups in the Pacific during the war, the navy badly needed to prove the Neptune's worth, especially for antisubmarine warfare. "A dramatic demonstration was needed to prove beyond question that the new patrol plane, its production representing a sizeable chunk of the Navy's skimpy peacetime budget, could do the job," writes a naval historian.[18] And with the AAF busily developing new ways of delivering nuclear weapons, the navy wanted to prove that it also had atomic capabilities. As another naval historian writes, the navy found itself in "a bureaucratic knife fight over roles/missions and ultimately, funding that turned on this critical capability."[19]

BILL IRVINE GAVE THE NAVY NO TIME TO ENJOY THE ACCOlades showered on its new plane. Two days after Christmas 1945, he set a transpacific record by flying against strong headwinds from Long Beach to Honolulu in a little less than nine and a half hours. He wasn't flying the *Dreamboat* but another B-29 called the *Challenger* that once was Lt. Gen. James "Jimmy" Doolittle's personal plane. Captain Saltzman again accompanied the flight. The *Challenger* didn't linger in Hawaii but took off the next morning for Guam and Luzon, the Philippines.

Irvine burst into the nation's and the navy's consciousness again in the early spring of 1946. He still wasn't flying the *Dreamboat* but a B-29 oddly named *Fluffy Fuz V*. Irvine didn't set out to establish another record; he simply needed to get to the Philippines, but all west-bound planes were temporarily grounded in Hawaii. "Typhoon conditions along the usual airway which runs from Honolulu to Kwajalein to Guam then to Manila initiated the record trip when Colonel Irvine found it impossible to return to his home base at Manila due to the unfavorable conditions."[20] Irvine naturally decided to fly straight to Manila instead.

The B-29 took off at 6:04 p.m. Friday evening, March 29, carrying ninety-seven hundred gallons of fuel. "The 'Fluffy Fuz' lumbered down the extra-long runways at Barbers Point Naval Air Station last night on the 5,000-mile great circle flight plotted to pass north of a typhoon off Guam," a mainland newspaper reported.[21] Captain Saltzman accompanied this flight, too, along with *Honolulu Advertiser* reporter Aaron "Buck" Buchwach, a young army veteran who later became the paper's editor in chief. "I hope we don't get near that Guam typhoon and drop one of those extra bomb bay gas tanks," crew chief M.Sgt. Gordon Fish told the newsman. Copilot Col. Beverly "Bev" Warren was more concerned about eating during the long flight. "If we haven't got more than two sandwiches each for the 20 hours," Warren jokingly warned Fish, "I'm going to get me a new crew chief."[22]

Fluffy Fuz V covered the five thousand miles to Manila in twenty-one hours and forty-six minutes. "Its progress from the time it smoothly negotiated a potentially dangerous takeoff at Barbers Point until it set down at the bomb-blasted key airport of the Army Air Forces in Manila was without incident," the *Advertiser* reported. Irvine had enough gas left that he could have continued to Tokyo. "It was a routine flight—the record was accidental," he said.[23] The colonel believed better communications facilities and landing strips in the central and western Pacific could significantly shorten air routes to Asia for military and commercial planes alike. "Wake island would serve best for traffic to Guam or the Philippines, while Marcus would be most practical for air travel to Japan or China," Irvine said. "In case of bad weather around either island, the other could be used as an alternate landing point."[24]

Irvine and the AAF kept up their assault on aviation records. "The following month, with time on his hands, he organized the Marathon Project on Guam. In five days, pilots and crews under his supervision set six new world's load-to-altitude records," the *Honolulu Star-*

Bulletin recounted.[25] Colonel Warren set one of the marks, carrying fifteen thousand kilograms (about thirty-three thousand pounds) above thirty-seven thousand feet, "an additional milestone in flying history."[26] The esoteric load-to-altitude records held little appeal to the general public but were important to helping the AAF determine how the Superfortress compared to aircraft from other countries. One record flight carried a dummy load of a thousand kilograms (twenty-two hundred pounds, or more than a ton) above forty-five thousand feet. "The Marathon Project is giving the AAF plenty of chance to learn what the B-29 can do; and the Superforts—they're keeping historians quite busy," a Boeing newsletter proclaimed.[27] Irvine directed ground operations and didn't fly any of the record-setting missions, but "you may be sure the man with the cigar is in the background, helping set the stage for his boys."[28]

STICKING WITH THE SORT OF AERIAL EXPLOIT GUARANteed to earn headlines, the navy countered *Fluffy Fuz V*'s transpacific flight exactly two months later with a record nonstop transcontinental hop by the new Neptune. The flight was east to west from Brooklyn to California against the prevailing headwinds and therefore slower than the records set in the opposite direction. Commander Davies left his desk in Washington to fly the P2V himself. He had come to appreciate the qualities of both the Neptune and the Lockheed vice president Mac Short. The combination, he said later, was an example of "an honest man . . . building an honest plane."[29] The Neptune he flew, bureau number (BuNo) 89082, was the first production model to roll off the assembly line following the two prototypes. Its call sign, Navy Nine-Zero-Eight-Two, derived from its bureau number.

Cdr. Eugene P. "Gene" Rankin was Davies's copilot. Rankin was a handsome former eagle scout and the son of a railroad engineer and a schoolteacher from Sapulpa, Oklahoma, near Tulsa. His parents had wanted him to be a doctor, but Rankin had other ambitions and so secured an appointment to the naval academy. At Annapolis, he was one of Davies's classmates and later an usher at his wedding. Upon receiving his commission, Rankin said he planned to stay in the navy until sixty-two, the maximum retirement age. "I should be an admiral by then," he added with a smile.[30] Following duty on the battleship USS *Colorado* and later a destroyer, Rankin shifted to the so-called brown shoe navy, where aviators wore brown footwear instead of the surface officers' black. He began flight training in 1940 and earned his coveted gold wings the following year, earlier than Davies did. During the war, Rankin earned medals and publicity as the commanding officer of Patrol Squadron 81 (VP-81), which in 1943–44 flew Consolidated PBY-5A Catalina patrol planes in the south Pacific.

The venerable "Cats" could take off and land from land and sea. First delivered to the navy in 1936, the PBY had retractable floats for water operations that folded up to become the wingtips during flight. "The overall effect was an aerodynamically clean aircraft far superior to previous flying boat designs," an aviation museum's website explains.[31] The versatile plane had a range of over twenty-three hundred miles and was good for multiple roles, including attack, reconnaissance, and air-sea rescue. The army also flew a version of the PBY in the Pacific that AAF Emergency Rescue Squadrons called the OA-10. Various allied nations flew the aircraft, too, and altogether Consolidated built nearly twenty-five hundred Catalinas.

The navy painted some of its Catalinas black for nighttime operations. Gene Rankin's VP-81 was among several acclaimed squadrons flying "Black Cats" in the south Pacific. Their chief mission, as *Flying* magazine observed in toxic language that was common during the war, was to "hunt down and destroy the pestiferous night-creeping J— barges that seek to reinforce and resupply the enemy."[32] Rankin later recalled that the high-winged, twin-engine Catalinas "took off at 85 knots, flew at 95 and landed at 85. But they got the job done."[33] Although predating the Hudson, Ventura, and Neptune, the Catalina would keep flying in civilian and military capacities worldwide for decades after the war.

Numerous newspapers published accounts of VP-81 working with PT boats during nighttime raids to harry Japanese forces. The commanding officer himself often flew missions. During six months of combat, the squadron, nicknamed "Rankin's Night Raiders," flew 1,777 sorties over 7,615 hours, 4,765 of them after dark. "They dumped 579,200 pounds of bombs on the J—, plastered them with 2,000,000 rounds of .30 and .50 caliber ammunition, after setting them up with 5,000 flares, and destroyed 34 barges and two merchant ships, besides damaging 23 additional barges," according to a navy radioman's hometown newspaper.[34] *Flying* proclaimed the squadron's record "truly impressive."[35]

Rankin's feat of getting all his raiders home alive was equally remarkable. "Flying the Navy's grand old Catalina flying boats from bases successfully in Guadalcanal, Munda and Bougainville, this famous Black Cat squadron has returned without losing a man or a plane in six months [of] operations," a North Carolina newspaper reported in 1944.[36] Rankin received several personal messages from Adm. William "Bull" Halsey, the commander of naval forces in the southern Pacific, and the Legion of Merit for outstanding leadership. He next reported to Washington, where he worked alongside Davies in the Bureau of Aeronautics. Davies couldn't have tabbed a more qualified copilot for the new P2V than his classmate and longtime pal Rankin.

THE PAIR BEGAN THEIR TRANSCONTINENTAL RECORD attempt on May 28, 1946. With them were navy lieutenants Melvin M. Pritchard and Frank P. Demont, Lockheed test pilot Harold S. Johnson, and flight test engineer Frank E. Osberg. "A ton of radar and radio apparatus flies with the navy's new land based patrol and search bomber, the Lockheed P2V Neptune," noted an Iowa newspaper.[37] Davies took off from Floyd Bennett Field at seven o'clock in the morning and turned westward from Brooklyn toward the plane's home nearly twenty-five hundred miles away in Burbank. The flight across the United States was uneventful until the Neptune hit turbulent spring weather over the Great Plains.

"Cmdr. Davies said the plane ran into a low-pressure trough between Wichita, Kan., and Amarillo, Tex., which forced it off course and up to 20,000 [feet] to get into satisfactory flying weather," the United Press reported. "Planned altitude for the flight was 10,000 feet."[38] Despite flying 150 miles off its anticipated track, the P2V still beat by thirty-nine minutes the previous record set in 1938 by Russian-born airpower advocate Alexander de Seversky. "The Navy's entry for atomic age bombing, the Lockheed Neptune, arrived here today twelve seconds after 1:23 P.M., Pacific standard time, making the trip from Brooklyn in 9 hours 23 minutes and 2 seconds, a new record," the *New York Times* reported.[39]

The navy welcomed the publicity, but with aviation records falling like tenpins to the AAF, it needed to do far more. Bill Irvine, after all, seemed to set a record every time he took off in the *Dreamboat* or any other B-29. A month before the Brooklyn-to-Burbank hop, a group of navy design officers in the Bureau of Aeronautics began planning a much grander Neptune flight. The officers first considered a globe-girdling nonstop flight. But that would have taken the P2V over Soviet territory, and souring international relations meant Moscow wouldn't grant permission. A sense of urgency to launch some other headline-grabbing flight reached the peak of the navy's chain of command. Admiral Nimitz, now the CNO, sent an ambitious memo in July to Secretary of the Navy James V. Forrestal.

"For the purpose of investigating means of extension of present patrol aircraft ranges, physiological limitations on patrol plane crew endurance and long-range navigation by pressure pattern methods," the memo said, "it is proposed to make a nonstop flight of a P2V-1 aircraft from Perth, Australia to Washington, D.C. with the possibility, weather permitting, of extending the flight to Bermuda."[40] A former naval aviator and author has observed, "Left unsaid in ADM Nimitz's memo was the fact that the intended route would exceed the distance record set the year before" by Colonel Irvine in the *Dreamboat*.[41] With rumors swirling that the AAF was planning something even more ambitious, Forrestal signed off on Nimitz's suggestion.

Davies's Neptune had a catchy new name by the time it headed for Western Australia two months later. The navy knew the army had a head start on making the next great long-distance record flight, so the Neptune's moniker arose from Aesop's fable about the tortoise and the hare. Although land based like the tortoise, the P2V belonged to the sea service and might rightly be considered a turtle. Commander Rankin, who had remained with the plane, explained that the nickname was a "subtle dig" at the army. With its many high-speed, record-setting planes, he said, the AAF "might look on our heavier patrol planes as the hare regarded the tortoise. But it's not always the fastest that gets there first."[42] Davies later confirmed the nickname's origins to a newspaper columnist: "The Army has a similar stunt called 'Project Marathon' and we remembered the turtle came out all right in the fabled marathon."[43]

The wife of a Lockheed engineer arranged for a cartoonist at Walt Disney Studios to draw an insignia for the plane. Soon a pipe-smoking male turtle wearing an officer's cap adorned the Neptune's nose. The reptile serenely pedaled a bicycle sprocket that whirled a propeller behind him, and a rabbit's foot on a chain streamed from the naval jacket he wore over his shell. Below him was printed *THE TURTLE*.

FIVE

Superfortress

THE B-29 SUPERFORTRESS WAS A GIGANTIC PLANE, NEARLY twenty-five feet longer and with a wingspan almost forty feet wider than the earlier B-17. Both bombers descended from the Boeing XB-15, a one-of-a-kind experimental plane begun as a design study in 1934. Mutt Irvine flew it in 1937.

"Lieut. Irvine flew out [to Seattle] last week to supervise the testing of the biggest air craft that flies and when he completes his work and formally approves the ship he will fly it back to Dayton for the government," the *St. Paul Phonograph* reported.[1] People later called the XB-15 Grandpappy because of the two bombers that followed it. Boeing submitted the first proposal for what became the B-29 before the war but didn't begin building it until several years later. "The Super Fortress is an all-metal monoplane, 98 feet long, 37 feet high, 141.2-foot wing spread, does 300 m.p.h. plus, gets up to 30,000 feet altitude, has multiple gun power turrets . . . and carries a far greater bomb load farther, faster and higher than any plane ever built," a newspaper reported in introducing the plane to the American public in June 1944.[2]

The B-29 was bigger and faster than the B-17, but the kinship was easy to spot in the long fuselage and high tail. The Superfort's most distinctive feature, though, was its bullet nose. Rather than having a cockpit jutting above the fuselage like the B-17, the B-29 had a smooth nose with windows set directly into the skin. The design was so futuristic that thirty-five years later filmmaker George Lucas would use it as the model for the *Millennium Falcon,* Han Solo's spacecraft in *Star Wars.*

The Superfortress was America's heaviest production plane, with many features and equipment built in to increase its range, bombload, and defenses. The normal crew of eleven included four gunners and a bombardier. A thirty-five-foot tunnel over the bomb bays connected the pressurized crew areas fore and aft. The plane was almost entirely electrically operated, and its guns in their turrets could be fired by remote control. Two R-3350 Cyclone 18s were mounted on each wing. Introduced by the Wright Aeronautical Corporation in 1937, the eighteen-cylinder, air-cooled, 3,350-cubic-inch radial engine provided from twenty-two hundred to thirty-five hundred horsepower, depending on the model.

The B-29 was the first major military aircraft to use the Cyclone 18. Various military and commercial planes relied on Cyclones later, including the P2V Neptune, the Consolidated Vultee B-32 Dominator, the Martin Mars flying boat, the Lockheed civilian Constellation transport, and even the single-engine Douglas A-1 Skyraider, first delivered in 1946 and still flying during the Vietnam War nearly thirty years later. "The power and long-range economy of the Cyclone 18 opened up new possibilities in aircraft design," Wright boasted in newspaper ads after the war. "Around it was developed the Boeing B-29, which enabled us to carry the Pacific campaign directly to the enemy's home islands from distant bases. With war's end, Cyclone 18s became available to commercial airlines."[3]

"The B-29 was an immensely complex aircraft, the most sophisticated, advanced bomber of World War II," aviation historian Walter J. Boyne wrote decades later. "Each of its major features might have taken five years to test under ordinary circumstances."[4] As with so much else, the Cyclone 18 was rushed into service too quickly and consequently was prone to failures and fires. The second B-29 ever built—still secret at the time and identified by the company only as "one of several Army bombers used by the Boeing Co. in its continuous flight-research program"—fell in flames over Seattle in February 1943.[5] The crash killed famed Boeing test pilot Ed Allen, the ten other men on board, twenty workers in the meatpacking plant the bomber crashed into, and a firefighter battling the ensuing blaze.

B-29 development continued nonetheless. The task of producing the plane fell to Boeing in Wichita, Kansas, but the plant wasn't prepared to handle it. "Wichita had been called upon to obtain tooling that did not exist," a modern article explains, "install them in a phantom factory, manufacture the world's most sophisticated bomber whose design and systems were constantly in flux, and build it with non-existent, unskilled labor that had no training or experience for such a task."[6] The AAF set aside normal testing procedures to get the bomber out to the Pacific as quickly as possible. Boeing in Renton, Washington; Bell Aircraft in Marietta, Georgia; and Glenn L. Martin Company at Bellevue, Nebraska, also set to work manufacturing B-29s. Simply training the workers was a huge task.

"The people there were untrained, plant facilities were practically nonexistent, and there were none of the conditions that would normally seem necessary for a successful industrial venture," an AAF general observed of the Marietta plant. "Believe it nor not, people who were employed to make aluminum planes had to be shown what a sheet of aluminum looked like."[7] Yet such workers built nearly four thousand B-29s before ending production in 1946. By the autumn of 1943, the Boeing plant workers in Wichita knew more than any other factory's about building the bomber. "Construction of the first bombers therefore was concentrated there under rush orders before all the 'bugs' could be eliminated," *Kansas History* recorded shortly after the war's end.[8]

"Superfortresses unready for battle were delivered to Kansas bases, where bombardment groups were poised for overseas," the *Saturday Evening Post* recalled. "Army mechanics at Salina, Pratt,

Walker and Great Bend [Kansas] tried to button up jobs left flapping."[9] General Arnold took an intense personal interest in getting the B-29 into the air. Late in the war, he delivered a brief speech to Superfort pilots heading for the Marianas. "Gentlemen, you can't win a war by circling over the control tower," he said. "Your Superforts were built to cross oceans and continents. It's up to you to get the miles out of your airplanes. Keep your ships out of the squirrel cage and fly them dry."[10]

First, however, the army had to get the planes out of the factories and on their way to the war. Arnold arrived in Kansas for an inspection in March 1944 only to learn that none of the B-29s scheduled to leave for India the next day was actually ready. The AAF chief detonated with the terrifying effect only a senior military commander can muster. His fury sparked what became known as the Battle of Kansas to get production back on track and to deploy 150 Superforts to the CBI theater by mid-April. "He exploded a string of 'impossible' orders that set phones jangling all over the country," according to the *Saturday Evening Post*.[11] Arnold appointed Brig. Gen. Bennett E. Meyers as the special project coordinator; Meyers in turn picked Bill Irvine as his deputy. Irvine had worked on the plane since its inception. "With Arnold's rage as their clout, Meyers and Irvine imposed order on the chaotic program," historian Boyne later wrote.[12]

Irvine was assigned to the AAF Materiel Command at Wright Field when the war began and was involved with engineering, inspection, industrial planning, and production control. "Recently he was in an accident and injured his spine so that it was necessary to wire the vertebrae together," one of Irvine's hometown newspapers reported in summer 1942, while also noting his promotion to colonel. "It was considered doubtful if he would ever be able to pilot a plane again."[13] Irvine mostly flew a desk for a while but soon was heavily involved in producing bombers and becoming so valuable that he was awarded the Legion of Merit. "His long engineering experience was of great value in determination of the proper types of aircraft to be procured and in guiding test programs that would bring forth new models and related equipment embodying the greatest military potential," the citation read in part.[14]

The colonel reported to AAF headquarters in Washington early in 1943, first as a special assistant for aircraft production and then in April 1944 as the chief of the Very Heavy Bomber Program. "In these jobs, he was deeply involved in the production and modification of the B-29 Super-Fortresses," an official biography says. "Among his tasks were those of getting materials and production facilities for these first atomic bomb carriers."[15] An aviation historian describes him as "the bluff, tireless man who did more to win the 'Battle of Kansas' than any other man."[16]

Irvine and other tough bird (full) colonels overran the Jayhawk state, looking to boost bomber production. The pace soon grew frantic. Army mechanics flew in from across the country, Boeing sent in six hundred experts, the AAF cut red tape and extended civilian overtime hours, workers donned wool-lined flight suits and worked outside when they had no room left in the hangars, and delicate instruments arrived in ambulances and spare parts and engines in empty bomb bays. The enormous effort eventually paid off as B-29s left for the CBI theater to take the war directly to the Japanese Empire.

Success came with a high price, however. Engine problems continued once the Superforts went into production. Their initial deployment to India "was marked by a number of crashes with five near Karachi in a two day period that was linked to the high temperatures in the India subcontinent, which caused engine failure."[17] Only intensive investigation and engineering work at Wright Field eventually resolved the failures. B-29s flying from bases in China meanwhile dropped their first bombs on the home islands of Japan in June 1944. The AAF began launching raids from Guam, Tinian, and Saipan later that year. Now flying with reliable engines, the Superfortress proved a tough plane that could absorb a lot of punishment. An Australian correspondent went along on the first daylight mission, which was flown by eighty-eight unescorted B-29s in late November.

"We bombed the city of Tokyo an hour ago. It was easy, almost too easy," the Aussie wrote from the plane, hoping to live to file his story. "But now we are fighting a bitter battle to get back to Saipan. The tail gunner is unconscious, with a wound in the head, we've lost altitude and our gasoline supplies are more than dangerously low."[18] The B-29 landed on little more than fumes, with the gunner dead in the plane—but it got back. Only one Superfort was lost on the mission.

The B-29s defended themselves well and shot down over nine hundred Japanese planes during the war. And the bombers could absorb phenomenal punishment. A Northrop P-61 Black Widow night fighter out for a daylight hop to calibrate its radar once stumbled across a crippled Dreamboat returning from a Tokyo raid. Half the bomber's nose was gone, its pilot was dead, and the wounded copilot was struggling without instruments to reach the Marianas. The Black Widow guided the plane over American-held Iwo Jima, where the surviving crew bailed out. Ground control then relayed orders to shoot down the derelict bomber. The Black Widow made four passes, firing all its machine gun and cannon ammunition, but the Superfort kept flying. It was a "ghost ship, a Flying Dutchman," the fighter pilot wrote. "In full view of 10,000 men on Iwo, an empty ship had all but beaten us. We had never heard of an aircraft absorbing such punishment." The bomber finally began an elegant spiral with its dead pilot still in his seat. "The 29 was not shot down, but it flew into the sea as if into a mirror."[19]

THE MONUMENTAL EFFORT TO DELIVER SUPERFORTRESSES to the CBI and the Pacific paid off. Not surprisingly, Bill Irvine remained in the middle of things, having been assigned to another new job by General Arnold. "Now, Irvine, see if you can make the stuff you pushed out function in the theatre," the AAF commander said.[20]

Irvine flew to the Marianas in October 1944 to become deputy chief of staff for Brig. Gen. Haywood "Possum" Hansell's Twenty-First Bomber Command. The colonel took charge of "supplying, maintaining and increasing the combat capabilities of B-29s that were striking Japan."[21] In AAF parlance, Irvine was the command's S-4. Gen. Curtis LeMay arrived from the Twentieth Bomber Command to replace Hansell in January 1945, a critical switch that led to the end of unsuccessful strategic bombing and to the start of firebombing of Japanese cities. As author Malcom Gladwell writes, "That change of command reverberates to this day."[22] Six months later, LeMay took over the entire Twentieth Air Force, composed of the Twenty-First and the newly arrived Twentieth Bomber Commands. Irvine was the oldest officer on LeMay's staff, and the general relied on him to keep hundreds of Superforts in the air.

LeMay was famously tough, abrasive, and efficient. Later, during peacetime, he headed the Strategic Air Command as a four-star "architect of strategic air power" who "insisted that the nation be willing to use nuclear weapons when necessary."[23] He grew increasingly conservative and controversial after retiring as the air force chief of staff in 1965. Lemay's published suggestion for defeating communist forces in Vietnam, for instance, was to "tell them frankly that they've got to draw in their horns and stop their aggression, or we're going to bomb them back into the stone age."[24] The former general further tarnished his reputation by joining Alabama segregationist George C. Wallace's presidential ticket in 1968.

But in 1945, LeMay was running the air campaign against Japan from the Marianas. Bill Irvine's biggest problem while serving under him was not the general but an extremely long and inefficient supply line centered on Hawaii. Everything funneled through the islands, causing frustrating delays. And LeMay, Gladwell adds, is "someone whose entire identity is about problem solving."[25]

Irvine tackled the job with typical energy, focus, and disregard for red tape. He worked back-channel magic to assemble a small fleet of C-54 Skymasters. Rather than wait weeks for spare parts, the colonel used the big transports to fly in the gear directly from a supply depot at Sacramento, California. Including stops in Hawaii and Kwajalein Atoll in the Marshalls, the supply runs were reduced to only two days. In addition, Irvine somehow arranged for a pair of navy ships to deliver equipment from the mainland directly to Guam, bypassing Hawaii. When even these arrangements didn't keep equipment flowing quickly enough, Irvine established a direct link to Materiel Command in the Midwest, where he had worked so long and still had many pals. "I set up a big antenna on Guam pointed at Dayton, Ohio," he said later, "and then sent a man over to Dayton to build a similar one. Every night we would order supplies. So every time we had a boat or a plane on the west coast we had stuff waiting on it."[26]

Despite the desperate finagling, getting B-29s into the air and headed toward Japan was always a last-minute dash. "At five o'clock in the morning, we might have fifty Superforts out of commission for lack of parts," LeMay wrote decades later. "At eight o'clock, Irvine's C-54s would arrive, and at nine o'clock, those fifty B-29s would take off on a mission. We ran it that close."[27]

No outsider really understood how Irvine got equipment delivered so quickly, and his general was careful not to ask him. "LeMay didn't care about ruffled feathers or protocol. He ignored the official protests and pointedly declined to ask Irvine any 'stupid questions.'"[28] The colonel's methods nonetheless aroused the ire of Lt. Gen. Robert C. Richardson Jr., commanding the central Pacific forces. LeMay managed to defuse tensions during the war, aware that Irvine expected to be court-martialed afterward. The judicial hammer never dropped following Japan's surrender, but LeMay was unable to deliver the brigadier general's star he thought Colonel Irvine richly deserved. Irvine, for his part, credited the AAF ground crews for keeping the Superforts flying. A journalist spoke to him in March 1945 after three hundred B-29s bombed industrial plants in Nagoya only forty-eight hours after a similar raid on Tokyo. "Irvine said his crews did twice as much work in half the time normally expected of them," newspapers reported. "'Proof of the quality of work accomplished in 36 hours is the fact that not one aircraft was lost, due to any material failure,' Irvine said."[29]

Awarded an oak leaf cluster for his Legion of Merit as LeMay's S-4, Irvine simply wouldn't stay on the ground. From mid-December

1944 through the end of the war, he periodically flew combat missions, gathering firsthand information that helped him boost maintenance and operational efficiency. During July and August 1945, he voluntarily flew on the first staging missions through Iwo Jima to Aomori and Kumagaya on the main island of Honshu. These flights earned him an Air Medal. "Despite the hazards of blind flying in adverse weather conditions, enemy fighters and anti-aircraft fire, Colonel Irvine displayed superior airmanship, remarkable coolness under attack, and courage and leadership," the citation read.[30] The colonel also received the Distinguished Service Medal, Silver Star, Distinguished Flying Cross, and Bronze Star—a glittering array of what airmen called fruit salad to pin on his uniform. "His gallant actions and dedicated devotion to duty, without regard for his own life, were in keeping with the highest traditions of military service and reflect great credit upon himself, his unit, and the United States Army Air Forces," read his Silver Star citation.[31]

Bev Warren also built an outstanding service record in the Marianas. The son of small-college educators, the Baylor University graduate spoke with "a soft southern drawl which belies the determination inherent in everything he does," an aviation magazine observed years later.[32] Warren worked on B-26 and B-29 bomber production earlier in the war before reporting as the deputy commander of the Nineteenth Bombardment Group. He got his name into stateside newspapers in May 1945 when as a lieutenant colonel he flew on strikes against six airfields on Kyushu. "He led 12 missions over Japan and participated in the series of five fire raids which did so much to destroy Japan's war industries," the *Honolulu Star-Bulletin* reported later.[33]

Irvine first noticed Warren for his work at the Martin bomber plant in Omaha and kept an eye on him while the good-natured Texan flew combat missions from the Marianas. "After quite a few runs, I thought I ought to get him out before he got all shot up," Irvine recalled.[34] Warren soon moved to LeMay's Twentieth Air Force as the deputy chief of staff for engineering. "During this period he was awarded the Legion of Merit for developing B-29 cruise control procedures which were later adopted as Air Force standards allowing an increase of bomb loads, and at the same time reducing fuel loading," his official biography says.[35] As soon as the war ended, Warren flew to Japan to assess airfields for use by American occupation forces.

Warren's experience flying B-29 raids and his research and engineering background led to his participation in Irvine's Marathon Project. Irvine expected to continue their teamwork and lean heavily on Warren during a grandly ambitious flight he planned for the *Dreamboat* later in 1946. As the AP reported of the Texan, Warren was "one of the army's experts at getting overloaded planes into the air."[36]

SIX

Distances

BILL IRVINE'S LATEST PLAN FOR THE *DREAMBOAT* WAS unprecedented and audacious. He proposed to fly the B-29 nonstop from Honolulu, Hawaii; across the top of the world; and on to Cairo, Egypt—or even beyond. The notion was almost laughable. "When I first considered the flight I said: 'Bill, you're crazy,'" Irvine told newsmen.[1] But his current boss, Lt. Gen. Ennis C. Whitehead, the commander of the AAF's Pacific Air Command, didn't think the idea was crazy at all. He and other top commanders were keeping a wary eye on the Russians, former wartime allies who were quickly becoming antagonists. Against the backdrop of what presidential adviser Bernard Baruch later called a cold war, Whitehead approved Irvine's crazy idea.

"The army air forces have planned this sensational maneuver as a practical demonstration to the world and to congress of what air transport can mean in the atomic age if there is another war," the *Honolulu Star-Bulletin* reported. The paper added that AAF officers worried that manpower and resources had become dangerously depleted by the rapid demobilization following the war. Irvine's transpolar flight would not only demonstrate what AAF bombers could do but also help "inspire congress to make the funds available for a continuing 'adequate' air force."[2]

Japanese fliers reportedly had flown an experimental aircraft more than ten thousand miles in July 1944. "The photographic plane was designated A-26 by Asahi Shimbun, a leading newspaper, which financed its design as a patriotic gesture," war correspondent Ted Wagner wrote a month after Japan's surrender. A Tachikawa Aircraft Company executive told him that the plane's range "would have permitted a flight from Tokyo via the Aleutians to New York and then to South America."[3] A British fighter shot down the only other A-26 over the Indian Ocean during a Tokyo-to-Berlin flight with several high Japanese officials on board.

United Press correspondent George McCadden now wrote in Honolulu that the Japanese claim of a ten-thousand-mile flight had "never been repudiated by the American experts who confiscated the plane" after the war and shipped it to Wright Field for testing. McCadden added that General Whitehead recently "told this writer that in his opinion, the Japanese claim appeared to be substantial, in that the sleek two-engine craft seemed capable of such a flight." But the A-26 pilots had made their flight secretly on a triangular course over Japanese-occupied Manchuria, thus setting no official record, whereas the *Dreamboat* "will attempt to span literally half the globe" across the Arctic in full view of the international public.[4]

America's commercial airlines showed interest in the Far North as well. "They want air routes over the top of the world," aviation author Douglas J. Ingells wrote. "Since they can't get an outright Government subsidy the experiments being carried out by the Army Air Forces are vitally important to the future of our airlines in their world planning." Ingells quoted a polar expert who thought the Arctic Circle had the potential to become "the cradle of our new air civilization."[5]

Several crews had flown over or near the geographic North Pole before the war, but none had done so easily. The first fliers credited with reaching the pole were Lt. Cdr. Richard Byrd and Chief Floyd Bennett, USN, while flying a Fokker trimotor monoplane in May

1926. Bennett later died of pneumonia while preparing for another expedition, and the airfield in Brooklyn now bore his name. A diary discovered in a university archive has since shed doubt on Byrd's claim of reaching the pole, although he and Bennett may have come close; the question has never definitively been resolved. "If Byrd did not succeed, historians of polar exploration said, the team aboard the dirigible Norge—the Norwegian polar explorer Roald Amundsen, the American Lincoln Ellsworth and the Italian Umberto Nobile—should be recognized as the first to fly over the North Pole," the *New York Times* reported seventy years later.[6]

Ironically, given the AAF's goals for the *Dreamboat*, Russian fliers were the first to cross the pole and continue to a destination beyond. "An over-the-Polar region air route between the United States and Russia, foreseen here as a possible short-cut to Europe and Asia, was aviation's newest pioneering project today," the Associated Press reported in July 1935. According to the Russian consulate in San Francisco, the purpose of the six-thousand-mile flight from Moscow to Oakland, California, was "to survey the shortest possible air route" between the two countries.[7] "Determined not to be left behind in the race to dominate the skies, Soviet experts for years had studied the polar route as the shortest and most logical way to reach North America," author and broadcaster Lowell Thomas and son Lowell Thomas Jr. later wrote.[8]

Pilot Sigizmund Levanevsky was an accomplished flier known as the Soviet Lindbergh. Copilot Georgi Baidukov and navigator Victor Levchenko accompanied him on the first attempt. Their long-range radio-equipped ANT-25, the brainchild of up-and-coming Soviet aircraft designer Andrei Tupolev, was one of only two in existence. "The Russian plane is a giant, low-wing monoplane equipped with a 950 horsepower motor, all Soviet built," the *San Francisco Examiner* reported. "It is a mystery ship, with unusually long, narrow wings, and a stubby fuselage. It is painted grey with black stripes and bears the identification 'UR S. S.-25.'"[9] The glider-like Russian plane had a wingspan of nearly 112 feet and, according to an aviation historian, "was equipped with the latest developments in blind flying instruments for its time, including an artificial horizon and a turn and bank indicator. It also featured an early gyromagnetic compass, a solar course indicator, and a radio transceiver with a range of up to 5,000km/3,107 miles."[10]

The three aviators planned a route somewhat east of a true great circle route over the North Pole. The route may have been planned in part to avoid Finland, which had declared independence from Russia in 1919. The ANT-25 took off in a heavy rainstorm on August 3, 1935, from a Moscow military airport, crossed the Russian coastline nearly nine hundred miles north of Moscow, and was flying over the icy Barents Sea when oil began leaking. The fliers couldn't fix the problem in the air, so Levanevsky made an emergency landing and reported the problem after eight hours of radio silence. Once back in the air again, he then flew southwest to Leningrad (St. Petersburg). "The failure of the flight caused bitter disappointment here," the AP reported from Moscow, "but the general reaction was a belief Levaneffsky [*sic*] was wise in turning back rather than risk a crash in the cold of the polar regions."[11] The unhappy pilot vowed never to fly another Tupolev aircraft.

The Soviets tried another polar hop two years later after several successful long-distance flights within the Soviet Union. Pilot Valery Chkalov flew the second ANT-25, which sported bright red wings and a silver fuselage. Baidukov again was the copilot with Alexander Belyakov as navigator. Soviet dictator Secretary-General Joseph Stalin kept close tabs on the ANT-25s and approved the latest flight following a meeting with the crew. The fliers again set out from Moscow with Oakland, California, and its seventy-two-hundred-foot runway, the world's-longest, as their destination.

The plan had been for both ANT-25s to make the flight, but engine maintenance kept one on the ground. Chkalov took off early Friday, June 18, 1937, his plane overloaded with fuel. Stalin stayed away from the airport, and the Soviet government used diplomatic pressure to stifle news of the flight, which it didn't announce for twenty-four hours. "The flight started secretly because of the failure of previous attempts to span the 6003 miles between Moscow and Oakland," the *Oakland Tribune* reported. "It was not until 18 hours after the take-off that word of the adventure, designed to test the feasibility of air transport across the Pole to the United States, became known."[12]

The Soviets flew north through the twenty-four hours of northern summer daylight, but the weather didn't remain clear for long. Their first scare came nine hours into the flight when the plane began icing up and vibrating violently. Climbing above the cloud layer into sunshine resolved the problem, but flying as high as twenty thousand feet also forced the men to begin using bottled oxygen. The fliers managed to stretch their eight-hour supply for more than ten. "We learned a great deal about flying conditions over the North Pole during that time," Chkalov said dryly later.[13]

It wasn't surprising that navigation got dicey as the crew flew over trackless ice fields far up in the Arctic. "Only the sun compass would resist the 'jitters' when all other instruments danced crazily under the influence of the magnetic area," the AP reported. "The device, developed for Admiral Richard E. Byrd's flights into the Antarctic, charts an unvarying course, shows a true North during travel along a given meridian."[14] Baidukov recalled, "Our compasses became especially sensitive when we were near the North Pole, the hands rotating like mad with the plane's slightest inclination—unlike the imperturbable gyros."[15] A second scare came when the plane's cooling system froze, causing the engine to overheat. With the crew's drinking water solid as well, the only unfrozen liquid left aboard was the urine samples the men had stowed in rubber containers for medical research. They used this unorthodox emergency coolant to save their flight. Four Soviet scientists based at a research station near the North Pole heard the plane pass over but said they couldn't see it because of clouds.

The Soviet fliers received navigational help in the Arctic from North American authorities. Thanks to a secret request three weeks earlier by the Soviet government, a radio network linked the ANT-

25 to American stations at Point Barrow, Nome, Fairbanks, Anchorage, Juneau, Ketchikan, and Seattle; a Canadian signal station at Fort Norman in the Northwest Territories; and Soviet stations at Anadyr and Khabarovsk in eastern Russia. A U.S. Army master sergeant who had flashed word of the crash that killed aviator Wiley Post and humorous Will Rogers near Point Barrow in August 1935 sat in his isolated radio shack now listening for the Soviets. "Never once in all manner of weather were we off our course or lost," the fliers said later. "At times ice coated the windows of our plane heavily, but we were never lost. . . . The excellent cooperation given by signal corps radio men was very wonderful, and it is appreciated by us."[16]

Late on their second day aloft, the crew passed over Pearse Island, British Columbia, on the border with southeastern Alaska, and continued down the coast. The fliers fell short of reaching California, however, and set down Sunday morning at Pearson Field on the U.S. Army post in Vancouver, Washington, after sixty-three hours and sixteen minutes aloft. "Heading straight down the coast, they first wandered about in miserable flying weather and flew 125 miles south of here to Eugene, Oregon," an AP dispatch explained. "With the weather getting steadily worse, they banked around and headed northward until they sighted the barracks airport."[17] Chkalov said the ANT-25's motor had been running perfectly. "But those darkening fog banks—it was useless."[18]

The exhausted oxygen supply rather than weather ended the run; otherwise, the fliers would have climbed above the thick seasonal fog ahead. "There was not a drop of oxygen left in the tanks," Chkalov said. He and his companions were exhausted, and Baidukov was also feverish, so "it was foolish to risk it after we had gone so far."[19] The plane was six hundred miles short of reaching Oakland, where some two thousand assembled spectators went home disappointed, but the crewmen were happy still to be breathing.

A handful of astonished soldiers greeted the fliers at the little army strip located across the Columbia River from Portland, Oregon. The doughboys quickly fetched a Russian-speaking Reserve

Officers Training Corps student who was training at the post to act as a translator. George Marshall, then a brigadier general, was the post commander. He took the aviators home with him for breakfast and beds as a civilian crowd began to gather around the plane. Their host's identity perplexed copilot Baidukov. "I don't understand," he said to Belyakov, "is he a general or a marshal?"[20] Baidukov couldn't know then that during the coming world war, Marshall would wear five stars as the U.S. Army's irreplaceable chief of staff.

News of the plane's safe landing was met in Moscow with "almost ecstatic joy by leaders of the Government, the Communist Party and by the people." Stalin sent the crew a cablegram: "Congratulate you warmly on your brilliant victory."[21] President Roosevelt also telegraphed congratulations to the Russian ambassador, who had secretly flown to California to greet the crew. "The skill and daring of the three Soviet airmen who have so brilliantly carried out this historic feat commands the highest praise," his message said. "Please convey to them my warmest congratulations."[22]

Vancouver was fifty-three hundred miles from Moscow, but Baidukov estimated that between dodging storms, avoiding mountain ranges, and retracing their route while searching for a safe place to land, the trio had flown nearly seven thousand. "Climbing to altitude and dodging storms cost the aviators many precious hours," the *San Francisco Examiner* reported. "Three hours were lost in getting around a storm in the Barents Sea."[23] The flight set no records but won widespread praise. The Russians thought they had flown "within 20 miles of the North Pole" and regretted they'd been unable to see the Soviet observation post in the Arctic.[24] That evening they flew as commercial passengers down to Oakland, where five thousand people turned out at the airport for a belated greeting. The fliers later toured the United States, crossed the Atlantic on the liner *Normandie*, toured Europe, and returned to banquets and parades in Moscow, where they were hailed as heroes of the Soviet Union. Their ANT-25 was dismantled, crated, and shipped back to Russia.

The Soviet flight immediately sparked speculation about what a transpolar route might mean for world aviation. Chkalov thought regular commercial routes eventually might follow the one he'd taken but the other way around—west to east, from the United States to Russia—to take advantage of prevailing winds. "The North Pole in years to come may be just a 'way station' on the USSR–USA air routes, aviation experts declared today," read a wire report in the *San Francisco Examiner*. The article added that already "ten regular commercial airlines, covering 7,500 miles, are now established in the Soviet Far North."[25] The United Press, in turn, quoted an observation by Australian polar explorer Sir Hubert Wilkins: "The distance between suitable landing places along the Polar routes is not very great, and is well within the capacity of modern, highly efficient engines."[26] Whatever else the future held, the Soviets promised that more transpolar flights would soon follow.

THE SOVIETS QUICKLY USED THEIR ONLY OTHER ANT-25 to make good on their assurances. "Just 12 days after Amelia Earhart disappeared dramatically in the Pacific while attempting an around-the-world flight, a trio of Russian aviators blazing a new polar route had to take an unexpected forced landing—in a Riverside County cow pasture," the *Los Angeles Times* recalled decades later.[27]

The second crew of pilot Mikhail Gromov, copilot Andrei Yumashev, and navigator Sergei Danilin took off at 3:22 Monday morning, July 12, from Schelkovo Airport near Moscow. Like the two earlier flights, they pointed their plane's nose north toward the pole, with their destination again Oakland or perhaps beyond. Gromov's flight was as long and challenging as the previous two, but his crew had better luck. "They passed over Rudolf Island, the last point of land between them and the north pole, at 2:01 P.M. Monday, then battled snow and cyclonic winds before reaching the pole at 7:14 P.M., a little less than a day from Moscow," the AP reported.[28]

The plane passed over the Soviet weather observation camp on an Arctic ice floe. Later it skirted a storm in the Canadian Rockies so rough that Gromov radioed for advice on how best to get out of it. Late Tuesday night, the ANT-25 asked a U.S. Army Signal Corps radio station in San Francisco for weather reports and information about the San Diego airport. At 1:15 a.m. local time on Wednesday, the fliers radioed they were about fifty miles north of the Bay Area and would pass Oakland. The plane then seemed to disappear. As the hours dragged silently on, both American and Soviet officials sent up search planes from California airports. The U.S. Coast Guard sent out an amphibious plane from San Diego a little before seven o'clock, while the Soviets chartered a series of flights around Bakersfield. But by then the missing plane was safely on the ground.

Gromov had flown down the coast as far as San Diego; then with fog stretching across his path, he had turned around and made for March Field, an army airfield in rural Riverside County. It seemed like the end of Chkalov's flight all over again. "The mountain range between San Diego and March Field was confusing to them, and they did not know just where they were when the gasoline tank sprung a leak," a report said.[29] The ANT-25 touched down in a pasture three miles west of San Jacinto and thirty-five miles short of March Field after more than sixty-two hours in the air.

The trio of aviators climbed out of their plane smiling "but almost tottering from weariness." Several ranchers had seen them land, and a lumber company employee rushed over to greet them. "The only reason we landed where we did was because if we had gone farther we would have left the limits of the United States and landed in Mexico," Gromov said. "We wanted to rest our ship on United States soil."[30] An army major in a fighter plane set down in an adjoining pasture ninety minutes later, and highway patrolmen and sheriff's deputies arrived to guard the Soviet plane. A car finally came to take the foreign fliers to March Field.

Unlike the effort a month earlier, Gromov's flight did set a long-distance nonstop record of slightly over sixty-three hundred miles. It topped the previous mark of over fifty-six hundred miles set by a pair of French fliers on a flight from New York to Syria four years earlier. Gromov sent a message to Moscow: "Proved feasibility of trans-Pole crossings, found magnetic disturbances. Radio communications O.K. with Russian stations. We reached Canada, then faded out. Plane performed excellently."[31]

Soviet newspapers and party organs hailed the latest ANT-25 flight as a rebuke to the country's enemies and critics. "Capitalism strangles and wrecks talents. It gives them no freedom for development," said the industrial publication *Industralizatsiu*. "Charles Lindbergh opened a new era in aviation by his flight. But he had to escape from his fatherland—escape from his glory."[32] With the Spanish Civil War raging and fascism rising in Europe, the Soviet Union issued a warning to unfriendly nations. "Our industry can produce as many planes as the country needs," the communist newspaper *Pravda* declared. "Let this be kept in mind by foreign enemies threatening war. Let them remember the distance between here and their capitals is much less than the distance to Portland, Oreg., or San Jacinto, Calif."[33]

Americans generally hailed the Soviet accomplishment. Los Angeles threw a parade, child star Shirley Temple treated the fliers to lunch, the mayor and governor threw them a dinner, and aircraft plants and movie studios gave them guided tours. The aviation and industrial contacts Gromov made during his stay proved invaluable once America and the Soviet Union were fighting side by side during World War II.

The Soviets meanwhile hinted that a third flight might soon visit the United States, perhaps one carrying passengers and cargo. What followed, however, was instead a tragedy. Sigizmund Levanevsky, the pilot of the failed transpolar attempt in 1935, vanished in August while attempting to fly a massive Bolkhovitinov DB-A heavy bomber from Moscow to Fairbanks and then on to Chicago and New York. Despite extensive searches by Alaskan and Canadian pilots, the wreckage and bodies of the crew were never found.

BRITISH FLIERS BROKE THE SOVIETS' RECORD THE FOLLOWING year. A flight of three single-engine Royal Air Force (RAF) Vickers Wellesley bombers took off early on Saturday, November

5, 1938, from Ismailia in northeastern Egypt, bound for northern Australia. "Never before has an attempt been made on the record by more than one machine," the *Sydney Sun* declared, "and the fact that the Vickers Wellesley bombers plan to fly the entire distance in formation, heightens an already exciting adventure."[34] A fourth bomber followed by easy stages.

The Wellesley incorporated an advanced internal geodetic structure championed by the airship and aircraft designer Barnes Wallis. During World War II, Wallis would design the bouncing bombs used in the famous Dambusters raid in the Ruhr Valley. The Wellesley's wings and fuselage drew their strength from a shell of interlacing structural members arranged along geodetic lines, which are the shortest distance between two points on a curved surface. (A great circle route across a globe is a geodetic line too.) The Wellesley was aviation's first all-geodetic aircraft, a technique Vickers used again on the Wellington, the RAF's primary bomber at the beginning of the war.

The three Wellesley bombers on the Egypt-to-Australia hop were part of what the RAF called its Long-Range Development Unit. They carried extra fuel tanks and special navigational equipment, with three-man crews instead of the usual pair. Four of the unit's Wellesleys flew a proving flight nonstop in July from RAF Cranwell in England to Ismailia and Shaibah, Iraq, forty-three hundred miles in thirty-two hours. Squadron Leader Richard Kellett, with six other officers and two sergeants, now commanded the trio of planes that would attempt the record. Much of the world followed their record-setting flight attempt to Australia. "The airmen, clad in heated suits with special satin linings, are flying in Vickers Wellesley bombers modified from the regular Air Force equipment," the *New York Times* reported.[35] The *New York Sun* gave readers the exotic-sounding route: "The flyers expected to adhere closely to the great circle route by way of Persia, across the Indian Ocean, Hyderabad, across the Bay of Bengal, Andaman Islands, Borneo, then across the Timor Sea to Port Darwin."[36]

The flight hit thunderstorms two hours out of Ismailia, and the three planes got separated in the clouds. Two rejoined later with the third never far away. Shore stations tracked the fliers' steady eastward progress at around 170 miles per hour. "They kept in constant radio touch with Singapore air base—but, as an officer remarked glumly, 'They were not a bit chatty.' Nevertheless, they could be heard exchanging messages among themselves."[37]

The second night out the Wellesleys crossed the Malay Peninsula five thousand miles from Egypt. "The storm clouds began to build up as night fell," navigator Flight Lt. Brian Burnett recalled. "We saw nothing of Borneo as we were in extremely heavy thunderstorms, rain and lightning for about six hours."[38] The three planes neared northern Australia after two days. "During the final 800 miles of their flight . . . the monsoon—strong seasonal winds—combined with tropical rain reduced their speed and increased gasoline consumption," the *New York Times* reported. "They had planned to fly inland to Cloncurry, Queensland, approximately 500 miles southeast of Darwin, but decided against the effort as they neared Darwin."[39]

The third plane ran low on gas and landed at Koepang (Kupang) on the island of Timor in the Dutch East Indies (today Indonesia). It refueled and quickly took off again in pursuit of the other two, having already beaten the Soviets' 1937 record by nearly three hundred miles. The leading pair pressed on toward Darwin, their sleepless crews managing to shave and straighten their uniforms before arriving. The two Wellesleys popped out of heavy clouds over the Northern Territory's capital the afternoon of Monday, November 7, 1938. The bombers peeled off right and left, and landed at Ross Smith Aerodrome three minutes apart at around 1:30, with Squadron Leader Kellett touching town first.

The two bombers had flown 7,162 miles nonstop and bested the Soviets by over 850 miles during two days and five minutes in the air. "Casually the personnel of the two Vickers-Wellesley bombers today stepped out of the machines after their magnificent record-breaking flight from Ismailia (Egypt)," the *Sydney Daily Telegraph* gushed. "Their R.A.F. uniforms were uncreased and spotless. They showed no fatigue."[40] The third plane landed safely at 5:30 p.m.

The Wellesley flights confirmed for British and Australian authorities a lesson the Soviets had learned from the ANT-25 expeditions: no major city anywhere was now beyond the range of another country's bombers. "The people of the United States should be more interested in the R.A.F. non-stop flight than any other nation," the *Daily Express* newspaper pronounced from London. "South America is only 2000 miles from Africa. Americans are now within the sphere of war terror."[41]

World War II began a year later in September 1939, ending quests for long-distance nonstop records. Kellett was captured after a bombing raid on Tobruk in 1942 and spent the rest of the war in a prisoner of war camp. Burnett served as a staff officer and after the war rose to the RAF rank of air chief marshal. Flight Sgt. (later Flight Lt.) Hector "Dolly" Gray, the wireless operator on Burnett's Wellesley, was captured by Japanese forces following the fall of Hong Kong, tortured during imprisonment, and executed in 1943. Gray posthumously received the George Cross for conspicuous gallantry in aiding fellow prisoners. The long-distance record the RAF fliers set in the Wellesleys was broken in November 1945 by Bill Irvine and his *Dreamboat* crew on their Guam-to-Washington flight.

SEVEN

Honolulu

HONOLULU WAS A MILITARY TOWN. PEOPLE THERE KNEW the *Dreamboat* was coming and why. The army made no secret of the fact that Bill Irvine's next goal for his record-setting B-29 was to fly from Oahu over the frozen Arctic to the Egyptian desert. Diamond Head to the North Pole to the Nile—the poetry and audacity of the thing fired the imagination. But the colonel had trouble getting his crew out to the starting line.

The *Dreamboat* got four new engines at Tinker Field in Oklahoma City before Irvine flew the B-29 to Seattle to prepare for the polar flight. The plane "rested quietly at Boeing field here today," the United Press reported on August 5, "awaiting completion of modifications for a proposed flight to Cairo."[1] The *Dreamboat* was still in the Northwest a week later, "undergoing repairs" that delayed its departure several more days.[2]

Irvine later wrote glowingly of Boeing's efforts in Seattle. The work included the "installation of new fuel-injection engines; the design and installation of a new streamline, light-weight empennage [tail assembly]; the new light-weight center-wing nylon tanks; the new large-capacity bomb-bay tanks which had been designed and built by Boeing-Wichita; the new oil transfer system; and finally, and most important of all, the fast-operating landing-gear mechanism," which retracted in only nine seconds and extended in fewer than three.[3] The *Dreamboat*'s six tires were later filled with lighter-than-air helium, saving weight for the Cairo flight. And they were inflated 50 percent more than normal to accommodate the enormous load and, in turn, required altering the wheel wells so the tires fit into them.

The preparations sounded impressive, but Irvine wrote privately in mid-August to Lt. Gen. Ira C. Eaker, the AAF's deputy commander in Washington, about "difficulties" in Seattle. "I personally think you have been wise in taking care of all the details to insure the safety of the crew in this long flight," Eaker replied. "That is always well for the success of the mission, in which there is a great deal of interest here."[4]

Capt. Ruth Saltzman was as intrigued by the Cairo flight as anyone on the *Dreamboat*'s regular crew. She "begged to go along, but Irvine refused," an air force historian writes. "He sent her to Washington to act as a liaison officer, and she spent $50 on telephone calls attempting to persuade him to take her along"—all in vain.[5] Irvine's men finally got the *Dreamboat* back into the air for a test flight on August 22. "Except for breaking in the new engines, which will be done between here and Hawaii, Col. C. S. (Bill) Irvine, holder of the world's non-stop flight record, was ready today for his Hawaii-Cairo, Egypt, flight over the north pole," the Associated Press reported.[6] Technical representative Richard "Dick" Snodgrass and flight engineer Lowell L. Houtchens from Boeing were making the hop to Hawaii too.

Rather than proceed directly to the islands, however, the *Dreamboat* first visited a cluster of air bases in Northern California. The B-29 winged from Seattle to Fairfield-Suisun Army Air Base (today Travis Air Force Base), south of Sacramento, touching down at 1:45 on Friday afternoon, August 23, two hours ahead of schedule. Irvine wanted to take off for Hawaii the next day. A spokesman said the AAF hoped the transpolar flight "will prove that the

United States has the men and equipment to fly the army to any part of the world."[7]

An unknown somebody—probably a staff officer in Tokyo or Washington—meanwhile changed the bomber's name to spotlight its parent command. The *Dreamboat* officially and briefly became the *Pacusa*; then just as briefly it was the *Pacusan*. Both derived from the acronym for Pacific Air Command, United States Army, created through the consolidation of five Pacific air forces in December 1945. "The final 'n' was tacked on just to round out a word," the AP explained. "It doesn't stand for anything."[8] The ground crew painted PACUSAN BOEING *Superfortress* onto the B-29's nose, making the first word twice the size of the other two. But the original nickname was more memorable and perfect for headlines, so common sense soon prevailed. The acronym morphed into an adjective, with the plane forever afterward being known as the *Pacusan Dreamboat*.

The B-29 no sooner arrived in California than personnel discovered a structural defect in its massive tail. Compounding the delay to correct it, red tape required the crew to undergo "normal overseas processing required of all personnel going to Hawaii"—never mind that all the AAF men were veterans of the Pacific war. "Some were being inoculated at the same time as the giant plane was being given a routine checkup," according to the army.[9] As problems accumulated, people in Hawaii began wondering when the plane would arrive. Irvine informed the AAF in Honolulu that he planned to make test hops at Fairfield on August 26, "after making minor repairs to fuel injection pumps in three of the B-29's four engines," the *Honolulu Star-Bulletin* reported. The crew would need more time to prepare for the Cairo hop once the *Dreamboat* reached the islands. "A 7th AAF spokesman said at least three or four days will be required for final preparations and 'after that it will be a matter of sweating out the weather.'"[10]

The *Dreamboat* went up for a four-hour test flight to check gas consumption. Irvine wasn't happy when he landed, and more delays followed. "We don't know if it's the weather or mechanical troubles that is delaying Colonel Irvine," an army public relations officer admitted. The AP noted that the colonel had said earlier that the success of the transpolar flight would be in the "hands of God and the weather."[11] The Honolulu visitors' bureau meanwhile arranged a reception for the *Dreamboat* crew whenever it arrived.

Irvine planned to fly the twenty-five hundred miles from the West Coast to Oahu at varying speeds and altitudes, continuing his search for the sweet spot that would provide the maximum range for the least fuel on a plane that burned a gallon per mile. "The Dreamboat will be the lightest B-29 ever to land at Hickam field or for that matter, anywhere else," the *Star-Bulletin* reported. "It has been stripped of all non-essential gears and is carrying only enough gasoline to get it to Honolulu."[12] The paper was wrong about the gas; the *Dreamboat* would carry a large fuel load of 10,500 gallons to simulate conditions on the Cairo flight. But Irvine and his crew first made a short hop from Fairfield to nearby McClellan Field in Sacramento. They finally departed for Hawaii from McClellan at 6:24 a.m. on Saturday, August 31, using more than a mile of runway to get their heavily laden bomber into the air.

A price tag was now affixed to the transpolar flight, with the AP describing it as "a $3,000,000 experiment to determine the feasibility of a global air force."[13] Irvine scoffed at the figure. He agreed that the *Dreamboat* incorporated expensive new instruments and equipment. "But because we are testing those things on the Dreamboat does not mean they should be figured in the cost of the flight," he argued. "They would have to be tested later, perhaps under less convenient circumstances."[14]

The flight to Honolulu was relatively untroubled. The *Dreamboat* had nearly eight thousand gallons of gas left during its approach to Oahu's south shore. Irvine wrangled by radio with Hickam Field, where the Air Transport Command feared that landing while so heavy might damage either the aircraft or the runway. Officials wanted Irvine to land instead at the naval air station at adjacent

John Rodgers Field (now Honolulu's international airport). But he prevailed and thumped down at Hickam at 2:30 p.m., ten hours and thirty-six minutes after leaving McClellan. (Local time in the territory differed by thirty minutes from other time zones.) The *Dreamboat* taxied up to the Hawaiian Air Depot, an AAF supply, repair, and modification center. Sally Claudy, an attractive member of Hickam's air communications service, climbed up into the cockpit with a bouquet of flowers and flashed a smile for pictures with Irvine and Bev Warren.

Irvine admitted he still wasn't "quite satisfied" with his plane's performance. A reporter asked about the troublesome fuel-injection pumps. "They're OK now," the colonel said. "The timing was off." But one of the Wright Cyclone 18s had "misbehaved" on the way from the mainland, and Irvine ordered an additional hundred-hour inspection for the bomber, plus more check rides before attempting the Cairo hop. "After that everything depends on the weather," he said. "We might be here three days—or it might be three weeks."[15] His crew favored the three-day scenario, but the *Dreamboat* faced an incredibly long flight, and the multiple weather systems between Honolulu and Egypt had to align before the bomber could fly safely through. "We figure this should be right educational," Irvine said wryly.[16]

Weather conditions near the Aleutians were especially critical. As on the earlier Guam-to-Washington flight, the B-29 carried no deicing gear to reduce the weight. "We simply can't afford to ice up," an unnamed crew member told the *Star-Bulletin*. "If we do we're finished."[17] The army expected the Cairo hop to take roughly forty-two hours, depending on tailwinds. An unnamed crewman (perhaps the same one) provided another succinct commentary: "If we don't make it in that time we don't make it."[18] September 2 was Labor Day, and the weather in Alaska was "somewhat confused." Based on the latest data, Irvine thought the *Dreamboat*'s takeoff "probably won't take place until next week."[19]

General Whitehead sent a long message from PACUSA headquarters in Tokyo wishing the crew good luck on their journey. "Your

10,000-mile flight over water, wasteland, ice and sand has been on the planning board of this command for months," he wrote. "It is no ordinary flight, but one which calls for extraordinary fortitude, skill and experience."[20] The message was mostly for public consumption and, while no doubt appreciated, was hardly news to Irvine or his men. They spent the holiday working on the *Dreamboat*.

1. A Boeing B-29 Superfortress landing at Guam during World War II. Courtesy NMUSAF.

2. Col. Clarence S. "Bill" Irvine, who began his career near the end of World War I, here a major general in the 1950s. Courtesy USAF.

3. General of the Army Henry H. "Hap" Arnold, who demanded that the U.S. Army Air Forces quickly build and deploy B-29s against Japan. Courtesy NMUSAF.

4. Capt. Ruth A. Saltzman, Colonel Irvine's trusted assistant and the only woman in the Twentieth Air Force, here in Seattle in 1945. Courtesy the Boeing Company.

5. WAVE Ens. Eloise English, who later married Cdr. Tom Davies, photographed in Washington DC during the summer of 1944. Courtesy NHHC.

6. Cdr. Tom Davies standing beside the P2V *Truculent Turtle*. Courtesy NARA.

7. The *Truculent Turtle* flying over Australia, September 1946, without its wingtip fuel tanks attached. Courtesy NHHC.

8. The Neptune's whimsical nose art, designed by an artist from the Walt Disney Studios. Courtesy NARA.

9. The seaplane tender USS *Rehoboth* (AVP-50) shortly after commissioning in 1944. Courtesy NHHC.

10. The *Truculent Turtle* and the R5D *Flying Workshop* in Western Australia before the *Turtle*'s record-breaking flight. Courtesy NARA.

11. (*above*) Sailors filling the *Truculent Turtle*'s wingtip fuel tanks before takeoff, RAAF Pearce, Australia. Courtesy NARA.

12. (*opposite top*) The *Truculent Turtle*'s pilots shaking hands after landing in Columbus, Ohio. *Left to right:* Roy Tabeling, Walter Reid, Eugene Rankin, and Tom Davies. Courtesy NARA.

13. (*opposite bottom*) The *Turtle*'s crew during their impromptu press conference at Columbus. *Seated:* Tom Davies. *Standing, left to right:* Walter Reid, Roy Tabeling, and Eugene Rankin. Courtesy NARA.

THE TURTLE

14. (*opposite top*) The *Turtle*'s fliers and their families at NAS Anacostia. *Left to right:* Roy Tabeling; Ruth, Carolyn, and Walter Reid; Tom and Eloise Davies; Virginia and Eugene Rankin. Courtesy NARA.

15. (*opposite bottom*) President Harry Truman congratulating the *Turtle*'s crew in the Rose Garden. *Left to right:* Walter Reid, Tom Davies, Harry Truman, Eugene Rankin, and Roy Tabeling. Courtesy NARA.

16. (*above*) Rear Adm. Tom Davies in 1970, beside a commemorative display with the *Truculent Turtle* behind, NAS Norfolk. The map is somewhat inaccurate, and the nose art differs somewhat from the original. Courtesy NHHC.

17. Colonel Irvine and crew at Bolling Field after their return from Cairo. *Left to right:* Richard Snodgrass, Edward Vasse, James Dale, Frank Shannon, James Brothers, Norman Hays, Gordon Fish, Beverly Warren, James Kerr, and Clarence Irvine. Author's collection.

18. The *Dreamboat* at Bolling Field, with the PACUSA insignia visible on the tail. Courtesy NARA.

EIGHT

Perth

SPRINGTIME WAS APPROACHING IN THE SOUTHERN HEMISPHERE when USS *Rehoboth* (AVP-50) reached Freemantle, Western Australia. The U.S. Navy seaplane tender tied up at the North Wharf, berth number 8, at eight o'clock on Thursday morning, September 5. Freemantle faces the Indian Ocean near the southwestern corner of the continent and provides port facilities for Perth, less than a dozen miles away up the Swan River. Honolulans wouldn't have blinked at anything less than a battleship, but humble *Rehoboth* caused a stir Down Under. "AMERICANS BACK," the *Perth West Australian* declared, above an article and photo of the ship secured to the wharf.[1]

The headline was a reference to the U.S. Navy's wartime presence in southwestern Australia. After a long retreat from the Philippines in early 1942, a history declares, Fremantle was the "most significant Allied submarine base in the Pacific after Pearl Harbor."[2] U.S., Dutch, and British submarines launched over four hundred patrols from the port. American naval aviators arrived there too. Patrol Wing 10 flew Catalinas from a seaplane base on the Swan at Crawley Bay from 1942 to mid-1944. The Aussies were grateful to the Americans for helping to defend their country and for taking the war to the Japanese.

Many Yanks came to love Perth in return. Residents grew accustomed to seeing their "lumbering PBY's at radio-mast height over the suburbs at dusk," the *West Australian* recalled after the wing's departure. "Since those days when Crawley Bay was a hub of training routine for pilots and the 'Black Cats' hovered over Perth in late afternoon, returning from long Indian Ocean patrols, many of the officers and men have distinguished themselves."[3]

Now American bluejackets were back, and vendors and visitors swarmed *Rehoboth*'s wharf. "Western Australia must be God's Country!" Lt. (JG) Edward N. Horner, the ship's doctor, wrote home before even stepping ashore. "The Aussies have been bending over backwards to do things for us. Engineers, medical officers, entertainers, chamber of commerce, reporters, and best of all, people with newspapers, fresh milk, and supplies have been coming and going all day."[4]

The tender had "accommodation for crews of seaplanes, and carries petrol for aircraft and facilities for carrying out minor repairs," the *West Australian* reported. "She is under the command of . . . K. M. Krieger, U.S.N., who is himself a naval air man."[5] No one in Australia knew exactly why *Rehoboth* had come, however. A Brisbane newspaper observed that the 310-foot tender was berthed at Fremantle on "a secret mission."[6] No matter why it was there, the Australians rolled out the welcome mat for the visiting Yanks.

USS *Rehoboth* bore the name of an idyllic Delaware seaside resort but wasn't a happy ship. The tender had smashed into a pier in Seattle soon after its commissioning at a Puget Sound shipyard in 1944. Two five-inch guns forward and one aft lent a menacing air, but *Rehoboth* never saw combat and rarely performed the tasks for which it was built. The ship transited the Panama Canal in April 1944 and shuttled men and equipment to ports in North Africa, Brazil, and Great Britain. The war ended before *Rehoboth* returned to the Pacific. The ship then spent months shuttling between battered ports in Japan, Korea, and China, seemingly without aim or purpose except to witness the devastation caused by modern bombing. Floating mines remained a danger, and marksmen on the deck occasionally blew them up with rifle fire. Port calls at Shanghai, with its poverty, crime, and sickness, hardly seemed less dangerous.

The ship's three hundred crewmen came to consider themselves the fleet's orphans with no meaningful role to play. Sailors rotating home in the spring of 1946 left their shipmates "really shorthanded," an ensign recalled decades later. "We never had adequate equipment and staff in the communications area for our job." He recalled their former commanding officer as a complex mixture of "Captain Queeg and Mr. Roberts."[7] Commander Krieger, the new skipper who brought them across the equator to Australia, had spent most of the war as the commanding officer of Naval Air Station (NAS) Minneapolis, Minnesota, and wasn't what the gobs called a blue-water sailor. But duty in Australia helped to revive sagging morale among *Rehoboth*'s remaining ten officers and 160 enlisted men.

"It was a wonderful thing to be in Australia, because we had been without fresh milk for a long time," Lieutenant Horner remembered forty-four years afterward. "We dined on Australian crayfish about four times a week and it was really a wonderful improvement on being in China."[8]

The tender's presence in Freemantle became less puzzling on September 18 when the P2V *Turtle* touched down at Guildford Airport, today Perth's international airport. Commanders Davies and Rankin and two other Neptune pilots hopped ninety-four hundred miles in four days across the Pacific through the Marshall Islands, over New Guinea, and across half of Australia. It was a long haul but not unprecedented. The navy couldn't have assembled a better crew to make the flight. Indeed, the *Turtle* may have been the only patrol plane ever officially crewed by three full commanders and a two-and-a-half-stripe lieutenant commander.

Cdr. Walter Shipstead Reid was the third member of the U.S. Naval Academy's class of 1937 to join the *Turtle*'s crew. Born in Edmonton, Alberta, of "blonde Norwegian Viking blood," according to his yearbook, Reid listed tiny Georgeville, Minnesota, as his hometown but considered Washington DC his real home.[9] Before the academy, he had served as a page in the U.S. Senate, no doubt thanks to his uncle U.S. senator Henrik Shipstead of Minnesota.

Tom Davies was an usher at his classmate's wedding in 1939. Reid served on the cruisers *Tuscaloosa* and *Wichita* and the destroyer USS *Satterlee* before entering flight school in 1940, earning his gold wings in June 1941. He also flew Catalinas during Atlantic antisubmarine patrols out of Panama and Bermuda. He then commanded aircraft maintenance squadrons in Australia, in the Admiralty Islands in the South Pacific Ocean, and in Norfolk, Virginia. Reid later served at the U.S. Naval Proving Ground at Dahlgren, Virginia, near Washington. He was the only member of the *Turtle*'s crew familiar with Perth, having served with the Catalinas at Crawley Bay during the war, and his wife would say with a laugh that "had something to do with his wanting to go on the flight."[10]

Lt. Cdr. Roy H. Tabeling Jr. of Jacksonville, Florida, was the plane's junior officer. He was several years younger than the others but no neophyte. Tabeling entered the navy's V-5 aviation training program at the University of Florida, graduating in 1942. Years later he said he loved flying any kind of plane, "including the crates they come in."[11] He flew a Black Cat in Commander Rankin's VP-81 and, while a lieutenant in May 1944, pulled off a spectacular rescue by picking up a downed AAF pilot during a violent storm off the northeast coast of Bougainville. "Landing in the quiet water near shore, Lt. Tabeling then taxied four miles through swells that often broke over the entire wing and fuselage," a squadron history records. Following directions radioed by a fighter escort overhead, Tabeling located the downed flier's life raft, "and the crew pulled the grinning pilot from the water. Then a long and extremely hazardous bouncing run, and the plane staggered into the air."[12]

The navy ordered the *Turtle*'s four aviators to Australia via dry official language barely hinting at anything extraordinary. "On or about 20 August 1946, you will proceed to the Lockheed Aircraft Corporation, Burbank California," Tabeling's orders read, "thence to Perth Australia, and to such other places as may be necessary for temporary duty in connection with experimental flights in an aircraft."[13] Three Lockheed civilians also went along: flight test engineer Frank Osberg, who had accompanied the earlier Brooklyn-to-Burbank record run; flight test inspector Stuart Sanford; and radio mechanic Ralph Plue. The first Neptune ever to fly Down Under, the *Turtle* came "straight from the Lockheed factory in Burbank, California, to Australia on a test flight," the *West Australian* informed readers. The newspaper added that the plane's "details are still on the secret list."

Davies told reporters at Guildford he was sore from sitting in the pilot's seat for nearly ten thousand miles across the Pacific. An overhead leak in the cockpit had let in a little rain, he added, but otherwise the *Turtle* had performed flawlessly. "She's been my baby for the last two years, almost from the start, so you can imagine that I'm very happy that she has performed so well."[14]

Armed guards from *Rehoboth* roped off and surrounded the *Turtle* at Guildford. By the next morning, news broke about why the plane had come all the way to Western Australia. "Perth–Seattle Flight Soon," declared a headline in the morning *Perth Daily News*. "Newest American long-range plane Neptune will leave Perth late this month on a flight to Seattle, Washington State," the article explained. "This was announced today from Washington by U.S. Navy spokesman Vice-Admiral Radford." When a *Daily News* reporter pressed Davies for more details, the pilot "said that his lips were sealed."[15]

Arthur W. Radford was far more eminent than the spokesman indicated in the article. He was the deputy chief of naval operations for air and in 1953 would become America's top military officer as the chairman of the Joint Chiefs of Staff. Radford wisely didn't mention Admiral Nimitz's suggested destination for the Neptune of Washington DC or possibly Bermuda. While perhaps surprising to anyone accustomed to reading two-dimensional printed maps, Seattle and Washington both lie very near the great circle route from Perth to Bermuda, making all three destinations feasible. The navy kept that tidbit to itself for now and expected the *Turtle* to leave Western Australia before the end of the month.

"Of particular importance is the fact that, with a relatively brief period of preparation, we have managed all arrangements necessary for launching a long-range patrol plane from a remote location of the world for a major flight," Radford told reporters.[16] Like Bill Irvine's B-29, the striking deep-blue Neptune had a longer nickname now, coined by an unsung navy public relations officer—the *Truculent Turtle*. In modern marketing parlance, both the *Truculent Turtle* and *Pacusan Dreamboat* had become brand names for their respective services.

Radford's statement that the P2V's flight was "intended primarily to test plane endurance, crew fatigue and new navigational methods and that any record set would be incidental" was more than a bit disingenuous.[17] Interservice competition remained fierce, with the *Washington Star* observing that it "would not be surprising if [the navy personnel] had their tongues in their cheeks when they made their avowals."[18] Radford added that Davies had "complete latitude" in deciding whether to keep flying once the *Turtle* reached Seattle. "He can continue after he reaches the west coast, just as he thinks best," the admiral said.[19] The navy wasn't publicly seeking a record, but nobody in blue would be unhappy if the P2V crew happened to set one.

The last piece of the Australia–United States expedition assembling at Perth arrived a day after the *Turtle*. It was a big four-engine navy transport designated the Douglas R5D. The AAF called its version the C-54 Skymaster—several had been in Bill Irvine's little "airline" in the Marianas—and civilian passengers knew the plane as the DC-4. Franklin Roosevelt had used a one-of-a-kind C-54 as his Flying White House, which the press corps nicknamed the *Sacred Cow*; President Truman used it now. Skymasters also were a familiar sight to travelers at Perth because Australian National Airways flew them six times a week across the country to Melbourne.

The R5D arrived at Guildford from Adelaide, South Australia, where it was the heaviest aircraft ever to land at the local airfield. The plane was informally called the *Flying Workshop* and carried a crew of seven led by Lt. Cdr. K. K. Kelley. The plane touched down at Perth the afternoon of September 19 with thirteen navy and Lockheed technicians aboard. Its mission was to support the *Turtle,* and within three hours, the techs began overhauling the Neptune's engines. USS *Rehoboth*'s aviation specialists and workshops assisted as well. The *West Australian* observed that Guildford's tarmac was now "a reminder of the days only a few years ago when United States Navy planes were operating from this State. There were 'Gob' caps, flying peak caps, sailors' uniforms, and flying jackets and again the place looked like a 'little American colony.'"[20]

Australia also figured in an earlier historic transpacific flight but as the destination rather than starting point. Former RAF pilots Capt. Charles Kingsford Smith and Capt. Charles T. P. Ulm—along with two former U.S. Navy aviators, navigator Harry W. Lyon Jr. and radioman James Warner—flew from Oakland to Honolulu to Fiji to Brisbane in May–June 1928 in a trimotor monoplane called *Southern Cross.* "Your brilliant, courageous pioneering has advanced the cause of aviation and strengthened the bonds between your commonwealth and our country," President Calvin Coolidge cabled Kingsford Smith.[21]

Perthians grew fascinated with Davies's plane, which they usually called the Neptune rather than the *Turtle.* The *West Australian* marveled that the navy would attempt to fly a patrol plane over nine thousand miles nonstop from Western Australia to the northwestern corner of America. "To the layman looking at the Neptune as it rests under armed guard on the tarmac at Guildford," the paper reported, "it seems incredible that a twin-engined aircraft could fly such an enormous distance; but it must possess powerful motors that are not over-thirsty for fuel."[22]

Coincidentally, Perth now also hosted a second famous aircraft, which arrived the afternoon of September 20. "Not more than 100 yards away from the United States Navy's latest long-range patrol plane, the Neptune, on the Guildford airport tarmac stands a R.A.F. Lancaster, a veteran and a record breaker," the *West Austra-*

lian reported. This was the bomber *Aries*, visiting the opposite side of the globe from a celebrated series of flights it had made during the war over the Arctic. Its crew had come to the Southern Hemisphere to update the Royal Australian Air Force (RAAF) on the latest developments in navigation, instruments, and equipment. The Lancaster also "has flown some 63,000 miles on normal navigation flights, trials and research," said RAF air commodore N. H. D'Aeth, commanding the international mission.[23] The *Aries* would soon return to Great Britain in stages via Melbourne, Darwin, Singapore, Ceylon (today Sri Lanka), Aden (Yemen), and Malta.

The *West Australian* published side-by-side articles about the *Aries* and the *Turtle* along with a pair of three-column photos that ran one above the other with a common caption. Aussies were so intrigued by the visiting planes that the official in charge of Guildford's Department of Civil Aviation cautioned them not to trespass; the airfield was open to the public but with limitations. "When it was necessary to restrict the public to a barricaded area—as it was at present with the Neptune and the Aries at Guildford—he asked that visitors should co-operate and not try to break the barricade," the newspaper reported.[24] Visitors weren't allowed in flying areas, including taxiways or runways. Tarmac control officers ensured that aviation buffs didn't wander into restricted spaces.

The navy expected the *Turtle* to begin its unique flight within a few days. A typhoon smashing into Guam with hundred-mile-an-hour winds, however, prompted serious concerns, since the Neptune was relying in part on weather data sent from the island. "If the meteorological station on Guam is not put in order it will seriously affect reports to be used on the flight," Davies said.[25] It was a very long way to Seattle, and in Burbank the P2V had been "especially modified by the removal of all armament and combat equipment and the installation of fuel tanks wherever they could be crammed," according to a history.[26] The process had taken three months.

Davies shared a few unclassified details about the plane with reporters in Perth. Some suspected that its transparent nose had been whitened to hide secret gear inside, but the commander assured them the measure was taken simply to screen the radio compass. The *Turtle*'s tires had flaps built into the sides that opened under pressure, he explained, so the wheels rotated when lowered, lessening friction and the danger of blowouts on landing. The P2V also had once-secret aerials that shed static electricity that interfered with radio reception, a vital improvement for a plane flying alone across the Pacific. The aviators also had new plastic earphones, molded to fit each pilot's ears, that Davies considered a huge improvement. "With normal phones, the band gradually burns into your head and your ears are pinned back," he said.[27]

The U.S. Navy and RAF crews announced a joint open house the weekend of September 21–22. Aviation buffs in Honolulu, a military town, would have been lucky to catch a glimpse of the *Pacusan Dreamboat* during a test flight through the blue Hawaiian sky. They certainly wouldn't have been admitted to Hickam Field for an up-close and personal look at the bomber. But in Perth, twenty thousand visitors descended on Guildford Airport by car, by bus, and on foot to see the *Turtle* and *Aries*.

"There was a constant flow of traffic to the airport yesterday afternoon," the *West Australian* reported. "It meant a long walk from the main road for those who travelled by bus, but few were disappointed on reaching the tarmac. There they were able to obtain an idea of the size and variety of modern aircraft."[28] The R5D support plane stood on display too. Late Sunday afternoon a little de Havilland Tiger Moth training biplane taxied up and parked next to the *Turtle*, providing a sharp contrast in size and technology. The *West Australian* ran a photo of Davies chatting on the tarmac beside the *Turtle* with Captain Krieger of *Rehoboth* and Robert Bailey, the Lockheed project engineer who had come to Perth on the R5D.

"Before the Neptune takes off on its long-distance record attempt it requires the full co-operation of the ground crew and those who are providing supplies," the paper noted. It described the effort by using the old wartime phrase "combined operations."[29]

NINE

Gremlins

BILL IRVINE HAD PORTRAYED A GERMAN ACE IN THE SILENT movie *Wings* but now resembled a character who'd spoken two memorable lines in *Casablanca*: "Waiting, waiting, waiting. I'll never get out of here."[1] But unlike the desperate refugee at Rick's Café Américain, Irvine didn't fear he would die in Honolulu. For one thing, he had faith in his crew. Except for his crack navigators, both borrowed from Wright Field, and M.Sgt. Edward G. Vasse, a radio operator who had joined the crew in Oklahoma City, everyone had flown on the *Dreamboat*'s earlier record-setting flights. "They were hand-picked, and I had my choice," Irvine said.[2] All their names and ranks were painted now on the B-29's nose.

The colonel also had settled on several key details for the Cairo flight. For one, the *Dreamboat* would take off for Egypt at 10:30 in the morning, so the bomber could reach the remote Aleutian Islands before sunset. Flying through the Arctic twilight still wouldn't be easy. "It will be too light to see the stars and too dark to see the sun," navigator Maj. James T. "Jim" Brothers explained. "Therefore, celestial navigation is out. There will be no reference or check points to look for. Whether we keep on our course depends upon the directional gyro."[3] Brothers said they didn't know how the polar forces would affect the gyro, but it didn't seem to bother him much.

Irvine also planned to take off from a makeshift thirteen-thousand-foot strip at adjacent Hickam and Rodgers Fields, starting his run on the latter. "The runways themselves are parallel, but the fields were joined by connecting the runway of one with a wide taxiway on the other to make a continuous take-off strip," a magazine later explained.[4] The colonel had settled on his initial course, too: head west out over the ocean before banking around to the northeast. The *Dreamboat* would "risk the necks of its crew rather than the lives of Honolulu citizens," the *Star-Bulletin* reported. "Although favorable winds would dictate a takeoff over the city of Honolulu, the heavily loaded B-29, carrying a tank car and a half load of gasoline, will turn its nose in the opposite direction."[5] Irvine remarked later that it really didn't make a lot of difference. "We could take off in the direction of Honolulu and still turn before we got over the city," he said. "It's better this way, however, because there are fewer obstacles."[6]

The AAF announced on September 5 that the flight would begin two days later, a Saturday, if the weather cooperated. USS *Rehoboth* had arrived at Freemantle earlier that day on the opposite side of the world, but no one in Honolulu was worried about the *Turtle,* which hadn't yet touched down in Perth. Irvine's crew and the journalists covering them nonetheless were all eager to see the B-29 get away.

"'Dreamboat,' a Superfort that flies with history in its wake, waited impatiently at Hickam Field today for the signal that would send it off on a 10,000-mile nonstop trans-polar flight Oahu to Cairo," wrote former marine combat correspondent Keyes Beech. The reporter, who later won a Pulitzer Prize during the Korean War, added that Colonel Irvine, "the Air Force's cigar-chewing record-smasher, keeps one eye on his Dreamboat and the other on weather reports."[7] One of Irvine's stogies came from navigator Major Brothers, whose wife, Ruth, had given birth to a boy, James T. Brothers Jr., the previous Sunday. ("Ted Jr., 7 pounds 9 ounces entered the world as you started around it," read the message from Dayton.)[8]

Irvine took the *Dreamboat* and the new papa up late Thursday for a ninety-minute flight over Oahu. Bev Warren took the controls during takeoff. Afterward Irvine pronounced the test hop "a very satisfactory test flight, with little work to be done."[9] Boeing technical expert Richard Snodgrass had made various engine checks and agreed, but Irvine thought they still needed a couple more test flights for safety. The AAF in Washington pushed the *Dreamboat*'s takeoff back one day and then a second.

With the era of Rosie the Riveter only recently passed, the *Star-Bulletin* published an article Friday featuring Polly Encke, a junior sheet metal worker at the air depot's matériel center. "For the past few days she has been busy doing some metal work on the Dreamboat, working along with the mechanics, aircraft engineers and other males who have learned to respect her ability," the paper reported. Married to an AAF master sergeant, Encke also worked on the *Fluffy Fuz V* before Irvine's record flight in that B-29. She loved working on the bombers. "I wouldn't think of doing anything else," she said.[10]

Encke probably hoped to see Irvine take the *Dreamboat* up again that day. "Today's test run will be for the purpose of checking the Dreamboat's cowl flap opening," the *Star-Bulletin* explained. "Thursday's was to determine the proper spark advance."[11] But the hop was pushed to Saturday, perhaps due to General Whitehead's arrival from Tokyo. Irvine's crew wore their dress uniforms for an informal briefing from the PACUSA commander in the Seventh Air Force's war room at Hickam. Whitehead dismissed any notion the transpolar flight was a stunt, declaring it instead a useful mission for testing new AAF technologies and operational techniques. "The flight is practicable . . . if I hadn't thought so, I would have stopped it a long time ago," the general said.[12]

But the news from the weather office was dismaying. Conditions along the route to Cairo looked so poor that the mission was pushed back again to Monday. Buck Buchwach from the *Advertiser*, Beech of the *Star-Bulletin*, and a third journalist went along on what was anticipated as the *Dreamboat*'s final test hop a few minutes before eleven o'clock on Saturday morning. The newshounds got a scare soon after takeoff when the bail-out bell rang in the rear compartment where they rode. Master Sergeant Vasse shouted for everyone to pull on a parachute while he rushed forward to find out the trouble. The noncom quickly returned to say Irvine had pushed the alarm when the men in the back hadn't responded to a call over the intercom. "Crew discipline, Col. Irvine called it, but it sent the cold shivers up our backs, even in the very hot pressure section," Buchwach wrote.[13]

The *Dreamboat* flew southeastward down the island chain at ten thousand feet and toward the big island of Hawaii. "R. B. Snodgrass, Boeing flight engineer and the only civilian in the nine man crew, filled several sheets of paper with figures which will be translated into ways and means of squeezing the last mile out of the Dreamboat's gasoline load," Beech wrote.[14] Irvine kept the bomber on autopilot for much of the flight, but Beech briefly got to take the controls from copilot Bev Warren's seat. Buchwach was up front, too, as the imposing Mauna Loa volcano rose through the clouds. Warren turned to his fiercely concentrating colonel and joked, "Say, boss, I'd just like to have you look up once in a while . . . that mountain's 13,000 feet high, and our altitude isn't quite that high, I'm a'feared."[15]

The *Dreamboat* returned to Oahu after three hours with only a quarter-inch adjustment needed on the cowl flap for engine number 4. A B-17 photo plane circled around to capture pictures of the B-29 with Diamond Head in the background. Once the Superfort was back on the ground, seventy-five air depot workers, many of them barefooted women, set about waxing and polishing its aluminum skin. The AAF thought a thick wax coating might mitigate icing up north and maybe add a smidgeon of speed as well. But Irvine and his crew wouldn't see any benefit right away since the weather forecast was still unfavorable, and the takeoff got pushed back again to Tuesday or Wednesday. The bad news hadn't yet reached the pilot's old friends and neighbors back in Nebraska.

"If Col. C. S. (Bill) Irvine doesn't establish a new record and amaze the aviation world tomorrow, there will be a lot of disappointed—and surprised—people in his old home town," a Lincoln newspaper reported.[16] But the weather remained lousy on Monday.

"The highs are where the lows should be and vice versa," Irvine said. "All we can do is wait." Meanwhile, everyone at Hickam was ready. The Seventh Air Force announced that eight P-51 Mustangs from the Fifteenth Fighter Group would escort the *Dreamboat* for at least the first 150 miles. A pair of B-17s would tag along for perhaps 800 miles in case of an emergency. But time was ticking away. "If the flight is delayed past Sept. 15," the *Advertiser* reported, "an entirely new plan must be drawn up for communications and navigation, because of the change in seasons and conditions."[17]

THE AAF'S AND THE NAVY'S LONG-DISTANCE FLIGHT PLANS both depended on three factors: man, machine, and weather. Both services had excellent crews that could hardly have been improved. Both aircraft had received extensive modifications and upgrades, with certain troublesome gear getting extra attention. Weather, however, literally was a force of nature; all anyone could do about it was to observe, analyze, and look for a safe path through thousands of miles of ceaselessly shifting winds, clouds, and temperatures. Each service had assembled a massive and in some ways unprecedented weather network, each tailored to its own requirements, to see its crew safely through its flight.

Col. Donald N. Yates in Washington headed the AAF's enormous Air Weather Service, a division of the ATC that the armed forces had created during the war. The service operated over two hundred weather stations around the world; their sizes ranged from a handful of people working in isolated outposts above the Arctic Circle to hundreds of personnel in large commands called centrals. The most important of the latter was the AAF's Master Analysis Central in the Pentagon. "The mainstay of a split-second communications network, using every modern device to transmit weather information in code, is the radio teletype, connecting land stations with airplanes in flight, ships at sea and far-flung battle fronts," a wartime article explained. "These tie Washington into weather centrals issuing shortrange forecasts for their various war theaters."[18]

A graduate of the U.S. Military Academy at West Point, a qualified army pilot, and a Clark Gable look-alike, Colonel Yates now commanded the whole operation. He held a master's degree in meteorology from the California Institute of Technology and early in the war spent seven months in Russia with a military mission coordinating weather information. Later in England, he earned medals from three Allied governments for helping select June 6, 1944, as the best time for the D-Day landings in Normandy. Less than two years later, the new Air Weather Service got a request for a feasibility study for the transpolar flight. Yates promised Irvine all the support he needed to plan and carry out such a flight on the *Dreamboat*.

For the B-29's transpolar hop, Yates wrote later, "almost every tool was brought into play—surface reports, upper air studies, reconnaissance flights, long-range forecasts from Washington Weather Central, short-range forecasts from Hawaii and other points along the route, and the world-wide weather communications net."[19] By September 1946, the AAF had shifted a special weather unit from California's San Joaquin Valley to Anchorage, Alaska, for gathering meteorological, communications, and navigation data in what the AAF called Operation Stork. The flights by the army's Fifty-Ninth Weather Reconnaissance Squadron were "the longest ever undertaken for weather reconnaissance and will be a part of the peacetime aid given by the Air Forces to commercial interests as well as for military use," the *Oakland Tribune* reported.[20] Twelve weather B-29s attached to the Fifty-Ninth plus two B-17s based at Anchorage would report conditions in Alaska and the Arctic for the *Dreamboat*. "But this was only a side job," a reporter wrote afterward. "Their main duty is straight work—routine observation."[21]

The Fifty-Third Weather Reconnaissance Squadron, based at Grenier Field in Manchester, New Hampshire, was flying from Newfoundland, Greenland, and Iceland to support the *Dreamboat* too.

The squadron's Flight B "operated under trying circumstances during the entire month [of September] while performing the best possible weather reconnaissance with available planes and personnel for the Dreamboat Project. . . . Last minute notices to resume missions for Dreamboat were routine."[22] Pilots had to abort one mission when an engine caught fire.

Capt. Arthur Yorra of Boston commanded the Thirty-First Weather Squadron at Hickam Field. He was also the *Dreamboat*'s project weather officer in Hawaii, charged with amassing data from all along the proposed route to Cairo. Captain Yorra eventually would wear a colonel's eagle. "The program was soon divided into two phases—preflight forecasting and in-flight reports," the *Honolulu Advertiser* reported. "The AAF weather central in Washington assumed the gigantic task of checking wind and weather conditions along the entire route and at possible landing sites."[23]

The army meteorologists faced a vastly complicated task in preparing forecasts for a flight over nearly ten thousand miles across some of the world's most inhospitable, inaccessible, and least understood terrain. A true great circle route from Honolulu to Cairo would have taken the *Dreamboat* over far northeastern Siberia, but that wasn't feasible in the deteriorating political climate. "All controversial territory will be avoided," the AAF said. Asked if the army had worried that the *Dreamboat* might be fired upon over Soviet-controlled areas in Asia and Eastern Europe, an army spokesman replied, "That's about it."[24]

Initial planning resulted in two Arctic routes that avoided Russian airspace. The first was from Honolulu to Sitka in southeastern Alaska; then over the Artic and Greenland to Iceland, London, and Foggia, Italy; and on to Cairo. The alternate route was from Honolulu to Dutch Harbor in the Aleutians—over twelve hundred miles west of Sitka—then a great circle route from there to London and onward. Both routes avoided Siberia during the first part of the flight as well as Soviet-controlled areas later behind what Winston Churchill recently had labeled the Iron Curtain. The AAF chose the course through Sitka. "Selection of the more southerly route . . . will permit the plane to fly over the weather instead of trying to go around it," the *Advertiser* explained.[25]

The Air Weather Service also identified four areas along the route where weather was most likely to be troublesome: the low-pressure systems in the Aleutians and Iceland, and the mountainous west coasts of Alaska and Greenland, where fliers faced what forecasters called orographic cloud buildups. The ideal scenario for the *Dreamboat* would be the lows in the Gulf of Alaska, far northern Canada, and the Norwegian Sea. But an army weather officer later wrote with masterful understatement that such a pattern "is not the normal situation."[26]

The Arctic posed the greatest unknowns and challenges for the *Dreamboat*. The Russians had much better weather capabilities in the Far North than did the United States. "On the Russian side there are well over 100 observation points above the Arctic Circle," from Siberia to the Barents Sea north of Norway, an American weather officer wrote in late 1944. "On our side, we find an area of 1500 miles in diameter in which we can plot just one weather symbol in the southern section." American pilots who visited the Russians during the war, the officer added, considered the Soviet weather service "the best in the world."[27] The AAF planned to attack the problem from the air rather than from the ground or atop an ice cap. A special unit from the Seventh Weather Group at Elmendorf Field, Anchorage, was busily gathering data over Alaska as Irvine waited impatiently in Hawaii.

B-17s equipped with weather instruments would fly ahead of the *Dreamboat* during the Cairo flight, gathering data on winds and conditions aloft. "B-29s of the 59th Weather Reconnaissance Squadron will fly from bases in Alaska to cover the Northern Pacific area," the *Advertiser* said. "Boeing B-17 Flying Fortresses based in Greenland will provide forecasts of North Atlantic regions."[28] The AAF also said that if something went wrong and the B-29 was forced down, radio stations along the route could establish the *Dreamboat*'s position via triangulation using direction-finding equipment. Air-sea rescue crews were on standby at Hickam Field, Ladd Field in Fairbanks, and Meeks Field at Reykjavík, Iceland.

The *Dreamboat* had an additional safety net created by a worldwide network of amateur radio operators, who are popularly known as hams. These civilians would keep in touch with the B-29 by Morse code and voice transmissions. Like the *Truculent Turtle*, the bomber's call sign was the last four digits of its serial number: Four-Zero-Six-One. The Alaska Communications System, operated by the army's Signal Corps, would keep a continuous watch from station WXE outside Anchorage. Radio stations in Canada and Greenland, along with maritime radio stations on both U.S. coasts, had promised to listen for Four-Zero-Six-One as well. Probably more people would be listening for messages from one plane than at any time since the 1937 disappearance of Amelia Earhart and Fred Noonan in their Lockheed Electra somewhere near Howland Island.

The man charged with keeping in touch with all these sources during the flight was Lt. Col. Frank J. Shannon, who in civilian life was a top Philadelphia radio engineer. Shannon had begun a remarkable military career during World War I, leaving the Marconi service to enlist as a navy wireless operator. Navy secretary Josephus Daniels commended the twenty-year-old for his actions in July 1918 when a German U-boat fired on the oil tanker USS *George G. Henry* and set it afire. Most of the crew took to the lifeboats, but Shannon and five others stayed on board and fought the blaze until help arrived. That was his *second* encounter with a U-boat. "A few weeks before the Henry caught fire, it was chased by a German submarine and badly damaged by shells. Shannon, acting in the capacity of wireless operator, attracted United States destroyers to the scene, which forced the U-boat to submerge and flee," the *Philadelphia Inquirer* reported.[29]

Following the "war to end all wars," the young hero returned to the City of Brotherly Love, where he was "associated with the radio broadcasting industry in this city since its infancy."[30] For nineteen years, Shannon worked as "the impeccable chief engineer of Philadelphia's Radio Station WCAU."[31] Despite having no military obligation at his age, the former navy chief radioman received an army commission as a captain in July 1942. He served with the Twentieth Air Force in the Marianas, where younger officers dubbed him "Pappy." Shannon afterward flew on the *Dreamboat*'s historic Guam-to-Washington and Burbank-to-Brooklyn flights, as well as on the *Challenger* for the Long Beach–to-Honolulu record. The army discharged Shannon in April 1946, but Irvine recalled him to duty especially for the Cairo flight. Attempting to fly the B-29 across the Arctic without Pappy aboard, the colonel said, was like "fielding a football team without a left end."[32]

Shannon's team of widely scattered radio amateurs would monitor the *Dreamboat* during the Cairo hop. The safety of the bomber crew also benefited from a heartening bit of interservice military cooperation. "The army and navy may not be able to agree on the merger, but the navy is lending a helping hand to the army air force for the Dreamboat's 10,000 mile non-stop flight from Oahu to Cairo," the *Star-Bulletin* reported.[33] The navy's direction-finding control station on Oahu would plot the bomber's bearings during the early hours of the flight and pass the information along to the AAF's communications office at Hickam.

In the meantime, army aerographers collected daily weather information at various points in Alaska and Greenland, plus Paris, Rome, and Cairo, for transmission to the Air Weather Service headquarters in Washington, which consolidated the data and relayed it by radio teletype to Hickam Field. Yorra and his weather team then prepared sixty-hour preflight forecasts—the latest innovation—with their maps indicating wind, weather, and icing levels along the *Dreamboat*'s planned path. Once the B-29 was airborne for Cairo, preflight forecasts would be "augmented with the latest up-to-the-minute weather reports from AAF weather stations along the route," the army newspaper *Stars and Stripes* reported. "These reports will be radioed to the Dreamboat at regular intervals throughout the flight."[34]

On Tuesday, Irvine briefed his men on ditching procedures and crew discipline; then he took them up for what he hoped was the final dress rehearsal. But on Wednesday, September 11, the whole ambitious plan abruptly skidded to a halt. The problem was a minute split in a thirty-three-by-sixty-four-inch fuel tank in the port wing. The Goodyear Tire & Rubber Company had specially built the tank, which the AAF had installed at Oklahoma City. It was made of a type of rubber called Pliofilm and held about four hundred gallons of gas. "A green dye in the plane's gasoline, which was added to permit detection of just such leaks enabled the crew to discover the split when they landed Tuesday," the *Advertiser* reported. "The wing was stained green from the slow dripping of the gasoline."[35]

The crew could do nothing about the problem in Honolulu. On Wednesday, Lowell Houtchens, the civilian Boeing technician and alternate *Dreamboat* crew member, flew with the ruptured tank aboard an ATC plane to Los Angeles. A Goodyear plant there would rebuild the tank, which Houtchens hoped to have back and reinstalled by the following Monday. "It would have been quicker to install another, new tank," the *Star-Bulletin* reported. "But the Dreamboat's pliofilm tanks, wider than the standard type, is [*sic*] one of only two sets in existence. The other set is in another plane." The other aircraft was an early version of the C-97, a bulbous military transport based on the B-29.

The *Star-Bulletin* wryly assured readers that Irvine's crew wasn't settling permanently in Hawaii or planning a Christmas at the Moana Hotel. Some philosopher observed that it was better to discover the vital tank problem on the ground. "Bosh," cracked Maj. James Robert "Bob" Kerr, the plane's administrative officer and resident wit. "If it had happened after we took off nobody would ever have known what happened!"[36]

Although of little consolation to the crew, the *Dreamboat* couldn't have departed anyway because the weather in the Arctic remained awful. Irvine spent Thursday morning testing further modifications to the four Wright engines, waiting for Houtchens to return from Los Angeles with the repaired fuel tank Friday. "The changes were made to aid the engines' cooling system, and along with final adjustments on the cowl flaps are expected to mean a difference of perhaps 300 to 400 gallons of gas during the 42-hour flight to Cairo," the *Advertiser* reported. The colonel hoped to be off the following week but was keeping an eye on low-pressure areas over Alaska and Iceland that he called the "two worst lows I've ever seen on a weather map."[37]

Things looked better Monday. Irvine scheduled another test hop with the repaired tank on board and held out the possibility of taking off for Cairo early Tuesday. But the troublesome imps that fliers and aircraft mechanics called gremlins during the war were working overtime. "The hard luck Superfort, Pacusan Dreamboat, ran into more gasoline tank trouble today," the *Star-Bulletin* reported. The test flight revealed a new leak in the opposite (starboard) wing tank. "Hawaiian air materiel area crews worked until a late hour last night in an effort to locate the leak, but without success."[38] Irvine had no choice but to send the second ruptured tank to the mainland for repair. And the *Dreamboat*'s troubles at Hickam were even more alarming than outsiders yet knew. "Once it [the plane] caught fire; Irvine and Gordon Fish, the crew chief, put out the fire with extinguishers. During a test hop a signal flare was accidentally discharged in the cockpit," Keyes Beech later reported. "'Nobody got hurt,' said a crew member, 'but there was sure one helluva scramble.'"[39]

The AAF indefinitely postponed the takeoff for Cairo. Irvine told journalists on Tuesday "that an overhaul of the plane's fuel system was necessary before the Superfortress is ready to depart from Hickam field." Adding to his misery, weather conditions in the five geographical zones had deteriorated from good to "pretty bad."[40] Worse yet, the *Truculent Turtle* had arrived in Perth, giving the *Dreamboat* competition for the long-distance record and the world's headlines. The news came as no surprise to the army, which had gotten wind of the navy's plans several months ago while ordering its lightweight gas tanks from Goodyear. The company had replied that it would try to meet the army's deadline, but it was busy with a similar rush job for the navy.

Irvine made appropriate remarks in Honolulu. "I think it is good for the progress of aviation that the Navy is attempting one sort of a flight while the Army Air Forces is attempting another," the *Dreamboat* commander said in a press release, and perhaps he even meant it. "Both flights should furnish extremely valuable operational information, since the routes flown, types of airplane used, and other aspects are entirely different."[41] The *Advertiser* considered the interservice competition healthy. "This type of Army-Navy rivalry is to be encouraged, because the true 'winner' of such competition is the American public," the paper said in an editorial. It added that the "results of the flights will mean better military aircraft . . . and improved commercial air travel for the future."[42]

But pressure from Washington to get the *Dreamboat* back into the air was growing. The resignation of Egypt's cabinet in late September prompted a crack from an *Advertiser* columnist: "Could it be that the members got tired of waiting to welcome the Dreamboat?"[43] As the newspaper later recalled, "Some people in Honolulu, while the plane was beset here with mechanical and weather difficulties . . . were beginning to become impatient with the Superfortress. They were beginning to mutter."[44] It seemed possible now that the *Turtle* might get away ahead of the *Dreamboat*. Word came down to silence the AAF's publicity machine until the B-29 unquestionably was ready for takeoff.

TEN

Navigators

ALL FOURTEEN PILOTS, NAVIGATORS, AND CREWMEN ON the *Dreamboat* and the *Truculent Turtle* were all-stars. Each needed to be top-notch, given the scale and audacity of the flights they were about to attempt. Mainstream newspapers made no mention, however, of the absence of an African American among the two crews. AAF units were still segregated in 1946, and the navy would not have its first Black aviator until 1948.

All four fliers on the Neptune had flown Catalinas long distances over water during the war, often during nighttime patrols. Each certainly recognized the hazards of flying over the vast Pacific. They surely knew, for example, that Honolulu's airport was named for Cdr. John Rodgers, a resourceful naval aviator given up for lost in 1925 after running out of fuel over the Pacific and ditching his PN-9 flying boat at sea. Rodgers and his crew raised improvised sails and reached Hawaii nine days later, but he was killed the following year in an air crash on the East Coast. Along with the rest of the world, the *Turtle*'s crew also knew about the 1937 disappearance of Amelia Earhart and Fred Noonan in the central Pacific, despite U.S. Coast Guard radiomen on the cutter *Itasca* having urgently worked to guide the fliers' Lockheed Electra to tiny Howland Island.

The navy aviators planned to take turns flying the *Turtle* while handling specific duties as well. Tom Davies was the command pilot, Gene Rankin was a copilot and engineering officer, and Walt Reid was also the navigator and in charge of the fuel-transfer station. Like Frank Shannon on the *Dreamboat*, Roy Tabeling was a ham operator, so he took charge of the communications gear on the flight deck. In addition to Reid, Davies and Rankin would stand navigation watches and shoot the stars with "two Bausch & Lomb 61-90-04 bubble sextants," according to the optics company.[1] Along with the sextants, the *Turtle* carried magnetic and gyro compasses, a radio automatic direction finder, and a radar altimeter. A ton of radar equipment had been removed to save weight, but the plane could navigate by homing in on radio waves, although the technique was of little use in the central Pacific.

Admiral Nimitz indicated in his July memo to Secretary Forrestal that in addition the pilots would try what was known as pressure pattern, or drift, navigation. The navy, the AAF, Pan American Airways (Pan Am), and Trans World Airlines all began investigating the technique in 1943. "In its essence, it is following the outer rims of low barometric pressure areas to take advantage of tail winds," *Aviation News* explained. "The lows are predictable in advance by weather forecasters and the plane is navigated to reap the benefit of the following winds."[2] But pressure pattern navigation depended on thoroughly understanding the meteorological conditions along the route. If successful, the technique added miles to an aircraft's flight by deviating from a great circle course but also increased efficiency to extend its overall range.

Except for the distances involved, the *Turtle*'s planned flight from Western Australia didn't differ greatly from an unusually long Black Cat patrol over the South Pacific Ocean during the war. The navy fliers themselves would have acknowledged that the challenges facing the AAF navigators aboard the Superfortress bound over the Arctic for Cairo were different and significantly more difficult than their own. Colonel Irvine consequently had recruited two outstanding men to guide the *Dreamboat* on the perilous journey: Majors Norman Hays and James Brothers.

Major Hays hailed from Grove, Oklahoma, about a hundred miles from Rankin's hometown of Sapulpa. Newspapers mistakenly listed his home as Seneca, Missouri, because his wife had grown up near there. After earning an engineering degree from Oklahoma Agricultural and Mechanical College (today Oklahoma State University), Hays entered the army in May 1940 and earned a commission as a navigator the following April. Excelling at prestigious navigation schools in Miami and an RAF center in Canada, he then spent eighteen months in combat zones in the Aleutians, Italy, and Saipan. Hays was awarded a Distinguished Flying Cross while with the Twentieth Air Force for flying as the lead navigator during a B-29 raid on a naval fueling station at Tokuyama in southwestern Japan. Despite heavy antiaircraft fire, the citation read, "a visual bomb run was made, and 75 per cent of the formation's explosives struck home within 1,000 feet of the aiming point. Vital oil tanks and installations were left a mass of wreckage, and more than half of the total roof area was destroyed."[3]

In June 1945, Hays flew with General LeMay on a record Guam-to-Hawaii-to-Washington flight. After hostilities ended, Hays flew on the October westbound Guam-to-Washington hop made by the three generals piloting B-29s. When Colonel Irvine nabbed him for the *Dreamboat* on July 1, 1946, Hays was the chief of the navigation instrument section at Wright Field.

Hays began his new assignment by reviewing available data from earlier polar flights. He later wrote that the key attributes needed in his job were intellectual ability, emotional stability, sharp perception, and good motor skills. "Navigators must be able to make sound judgments, to deal correctly with sequential details and to organize efficient operation plans," Hays added.[4] The major possessed all these key abilities. A journalist later wrote that the navigator "looked like a kindergarten truant smoking a big black cigar," but probably no one knew more about guiding a B-29 on a hazardous mission.[5]

Major Brothers of Fountain City, Tennessee, outside Knoxville, was no mere backup for Hays but a decorated navigator in his own right. A former student at the University of Tennessee, he looked "more like a football player than an airman."[6] Brothers began the war as a private and noncombat engineer in the European theater. After returning stateside and earning a commission, he shipped out again to the Fifteenth Air Force as a navigator on Consolidated B-24 Liberators flying from Italy. The Tennessean flew fifty-one bombing missions during seventeen months overseas.

Brothers's plane returned from one mission with the copilot mortally wounded. During another mission over Hungary, his Liberator was hit by flak that stopped both engines on one wing. "We were barely able to get over the mountains," he remembered. "We followed the valleys and luckily met no enemy fighters. We reached base an hour after the others. They had given us up for lost." Brothers ended the war with an Air Medal with five clusters and a Distinguished Flying Cross with one cluster.

Coming home, the major lucked into a ride on a Skymaster with Gen. Mark Clark, the former commander of the Fifth Army in Italy. A crowd of forty thousand people met the general at the Chicago airport, and afterward Brothers and others on the plane rode in a ticker tape parade downtown and were introduced from the speakers' platform. "Most of us naturally just took a bow," he said.[7] Now the major was in Honolulu while his wife and newborn son were in Ohio, waiting for him to fly across the Arctic Circle and on and on to the ancient home of the pharaohs. Hays later wrote that he and Brothers were both "plenty worried. . . . There are strange things, almost phenomenal, about Arctic flying."[8]

NAVIGATORS FLYING OVER THE ARCTIC HAD TO CALCULATE for tailwinds, headwinds, mountain ranges on both sides of the vast northern ice floes, gaps in radio communications and in the availability of weather data, and more. An additional factor was the quirky nature of the magnetic North Pole, which wanders about like a lost explorer. The Associated Press later noted that scientists knew about the pole's mysterious shifts long before the *Dreamboat*'s flight but didn't understand the reasons for them. "In fact they estimated its course rather regularly throughout its 200 or more miles of movement since 1904, when Roald Amundsen spent a winter charting its location."[9]

In its coverage of Valery Chkalov's flight over the Arctic to Vancouver, Washington, in 1937, the AP referred to the Far North's "magnetic jungle, where the compass, key navigation instrument, pointed at almost everything but the North Pole."[10] In fact, a compass doesn't point toward the geographical pole but toward the magnetic pole, which usually is a considerable distance to the south and, during the twenty-first century, is moving rapidly toward Siberia. "The north magnetic pole is restless," the *New York Times* explained in 2019 when the point was especially nomadic. Early explorers' hand-drawn maps had shown the magnetic North Pole wandering the Canadian Arctic Archipelago. "Then around 1860, it took a sharp turn and bee-lined toward Siberia. Since then, the pole has traveled nearly 1,500 miles and was most recently found in the middle of the Arctic Ocean, still en route to Russia."[11]

Popular Mechanics simplifies the cause of the pole's wandering: "Two writhing lobes of magnetic force are duking it out near Earth's core."[12] In October 2017, the magnetic pole crossed the international date line less than 250 miles from the geographic pole. This erratic movement not only was a problem for navigators Hays and Brothers in 1946 but has proved so for airport administrators today as well.

Runways are numbered according to their magnetic headings, rounded up or down to the nearest 10 degrees. Runway 19L in Wichita, Kansas, for instance, is 190 degrees off north, with *L* or *R* signifying left or right for parallel runways. When the position of magnetic north shifts far enough, runway designations must shift too. "And so, any given year, it's likely that at least one or two airports will have to break out the white paint," *Wired* notes, adding that "it takes a decent amount of movement to trigger a change, and not every airport hits that point at the same time."[13] The runway numbers at Oakland International Airport in 2013, for example, shifted from 27 and 29 to 28 and 30—a change that might have amused Valery Chkalov and his companions who had landed there after their transpolar flight from Russia in 1937.

The wandering magnetic pole wasn't the only challenge facing the *Dreamboat*'s navigators. The plane would fly over the polar region at dawn, leaving Hays and Brothers without stars to help them navigate. Radio and radar fixes that far north provided no help either, and the B-29's long-range navigation gear had been removed to reduce weight, although the autopilot and some standard compasses remained on board. The *Dreamboat* also carried a small, new six-pound directional gyro developed by the Sperry Gyroscope Company. The device spun at twenty-eight thousand revolutions per minute and indicated deviations from a given course. Having bought twenty of the units, the AAF tested them extensively in a lab at Wright Field. The two most accurate units went through more testing, and the better one was then installed on the *Dreamboat*. The instrument was housed in a cage and disconnected from the plane's electrical system until needed during the flight.

Various cross-country hops and the long flight out to Oahu from the mainland gave the navigators opportunities for further testing. They would rely on their gyro's "little yellow line" while flying between the seventy-second parallel going north and the seventy-fifth parallel coming back south.[14] At the top of the world, the planet tends to turn from under a gyro, however, causing navigational errors if uncorrected. Hays and Brothers would compensate with a technique called precessing, with the necessary precessions already calculated and entered into the gyro.

In addition, the B-29 was equipped with a device whose name might have come from a Buck Rogers comic strip or movie serial—the Pioneer gyro fluxgate compass. Developed by Bendix Aviation during the war, its main purpose was to display steady readings during dives, climbs, turns, and evasive maneuvers that can send magnetic compasses haywire. "The sluggish needle is replaced by a fixed coil system in which actuating currents cleverly combine with energy generated by the earth's magnetic field," a Scripps Howard writer explained in 1943. "The fixed coil system is maintained absolutely horizontal by a gyro stabilizer, a simpler requirement which can be accomplished within aircraft weight and space limits."[15]

The *New York Times* described the fluxgate compass when the device was installed on B-17s bombing Nazi Germany: "To remove it as far from magnetic and electrical disturbances in the plane, the compass is mounted far out on one wing and connected to dials in the navigator's and pilot's compartments. 'It is just a compass with all the errors taken out,' one navigator explained."[16] The fluxgate compass was similar in concept to the gyromagnetic compass installed on the Soviet ANT-25, with the AP noting that the new device "keeps its accuracy nearer to the north pole than the conventional magnetic compass."[17] Before the Cairo flight, however, technicians removed the *Dreamboat*'s standard fluxgate compass. Hays relied instead on a smaller fluxgate that was part of the automatic pilot system, which he considered equally sensitive and as accurate as the standard fluxgate compass.

Like sailors for millennia before them, Irvine's crew also could navigate by checking the positions of celestial bodies on a clear night. The technique depended mainly on fifty-five stars—twenty-seven in the northern sky, twenty-eight in the southern—plus the planets Venus, Mars, Jupiter, and Saturn. "In polar flights celestial navigation is exceptionally valuable because of the crippling effect upon the magnetic compass and radar of nearness to the magnetic pole and unpredictable magnetic disturbances," *Nature Magazine* later explained.[18]

Majors Hays and Brothers had a standard astrocompass Mark II on board to determine the direction of the plane through sightings on such bodies. They also had two sextants for comparison fixes and a spare astrocompass in case anyone crawling through the narrow tunnel connecting the B-29's forward and aft compartments accidentally knocked the first astrocompass from its mount. In addition, the two navigators had a new celestial computer to check the plane's drift. This device provided an inverse stereographic projection onto a glass grid of the area the plane was passing over—if the sun was visible when the crew needed it. Having a celestial computer meant they didn't need to carry bulky navigation tables. Plus the plane had a low-altitude refraction correction table developed by the navy at its hydrographic office in Washington, a small example of interservice cooperation.

George McCadden of the United Press in Honolulu offered military-minded readers as much background on the *Dreamboat*'s navigators and their equipment as the AAF would permit. Lacking any detailed charts of the Arctic, the journalist wrote, Hays had created his own hand-drawn grid for the region. A guideline made across the grid would indicate his course over the top of the world. "His navigation problem is boiled down to 'staying on the beam' with this grid line," McCadden explained. "It will be the first attempt to make what he termed a 'straight line grid crossing' of the arctic." Having explained his gear and techniques to the journalist, Hays added, "Now I'm gonna see if the system works."[19]

THE RAF, RATHER THAN THE AAF OR THE SOVIETS, HAD the most recent experience in transpolar navigation—coincidentally in the Lancaster bomber that was the *Turtle*'s recent companion at Perth. "Lancaster PD328, named Aries, was a very special machine," writes a Canadian military historian. "Built as a standard Mark I bomber, it had completed a round-the-world flight in 1944 as part of RAF preparations to deploy heavy bombers to the Pacific theatre once Germany had been defeated. . . . In April 1945 the aircraft was modified extensively."[20] Built by Metropolitan-Vickers rather than the principal manufacturer, A. V. Roe & Company (Avro), the four-engine *Aries* then made a series of flights in the Far North in May 1945 shortly after Victory in Europe Day.

"During the closing phases of the Hitler war, the R.A.F. Empire

Air Navigation School at Shawbury had become greatly interested in polar aviation, and went ahead with plans for exploratory flights without waiting until peace had been achieved," aviation pioneer and author John Grierson wrote nearly two decades later.[21] Unlike nearly all other wartime missions, the expedition was widely publicized within Great Britain and English-speaking Allied nations. The Air Ministry announced in advance that the *Aries* was flying north "to study navigation in the conditions imposed by Polar flying, to examine the behavior of compasses and Radar installations, and to collect magnetic and meteorological data."[22]

Wing Commander David C. McKinley was the *Aries*'s pilot and perhaps the RAF officer best qualified to take the Lancaster north. In July 1941, five months before the United States entered the war, McKinley flew American presidential adviser Harry Hopkins in a Catalina from Scotland to Russia, where Hopkins conferred with Soviet leader Stalin a month after the German Army invaded the country. The hazardous, roundabout flights took more than twenty hours each way. McKinley later served as the chief instructor at the RAF's central navigation school and in 1944 piloted the *Aries* on its world tour. In 1945 his mission over the Arctic constituted "the first airborne comprehensive scientific investigations of the area and would represent the first British flight to the pole."[23]

The Lancaster on its Arctic flight was much like the later *Dreamboat*: stripped of weapons and all other dispensable equipment; its bomb bays and other spaces stuffed with additional fuel tanks, enough altogether for nearly four thousand gallons of fuel; and its aluminum fuselage shining and bare to save the weight of a coating of paint. The *Aries* carried multiple research and navigational gear and was unheated to avoid misting the windows and draining electrical power. The lack of heating took a toll on the crew, however. "Their efficiency was reduced both by the fall in body temperature and by the extra exertion and discomfort caused by wearing bulky clothing," the plane's medical observer wrote.[24]

The *Aries* left RAF Shawbury in Shropshire, England, on May 10,

1945, for Meeks Field in Iceland. Wing Cdr. Kenneth C. Maclure, Royal Canadian Air Force, was aboard to gather meteorological, radar, and special data. The flight plan to the geographical North Pole, he later wrote, was to have the sun "straight ahead as we bore down on the Pole, and sextant acceleration errors would be reduced to a minimum. It was at this stage that we had the novel experience of flying due north at midnight, with the sun straight ahead and rising in the sky."[25]

The bomber's eleven-man crew had only a five-day window in the north to complete their Arctic missions. They fully experienced the magnetic jungle and the wandering magnetic pole mentioned by the Soviet aviators after their earlier flights across the north. The RAF crew's first attempt to fly over the geographical pole on May 16 was turned back by low clouds and ice. They took off again only two hours later. "In their arctic flying suits, the British-Canadian crew of the converted Lancaster bomber looked like men from Mars," *Time* magazine told American readers. This time the *Aries* reached the pole a little past 2:00 a.m. on May 17. The fliers circled for ninety minutes and completed one "trip around the world in 76 seconds by tightly banking around the Pole," after which the mission's doctor tossed out a Union Jack and a bottle of beer to celebrate.[26]

The bomber crew started for the magnetic pole the following day, but electrical troubles forced a twelve-hundred-mile diversion to Goose Bay, Labrador. Steering mainly by astrocompass rather than by magnetic compasses, they finally reached the magnetic North Pole the next day before landing in Montreal. Finally, to make a practical test of their findings in the Far North, the crew flew the *Aries* west by easy stages to Whitehorse in the Yukon on May 22. Three days later, they flew without difficulties nearly forty-two hundred miles over the Arctic home to England. "The flight was uneventful but confirmed the hopeless inaccuracy of charted magnetic data throughout the whole polar area and of height data for Greenland."[27] Altogether the Lancaster flew nearly twenty-four thousand miles during more than a hundred hours of research flying, an impressive feat for a country still at war with Japan.

"No fewer than eleven compasses were aboard the machine," the aviation correspondent for London's *Guardian* newspaper reported, "and the captain . . . told me that most of these 'did not seem to know what to do' when the Aries was flying over the Svendsrup [*sic*] Islands, which is where the magnetic pole is believed to be."[28] Situated in the Arctic Ocean and bearing the name of a Norwegian explorer, the remote Sverdrup archipelago lies within the Northwest Territories of Canada; the nearby Boothia Peninsula to the south was the last charted position of the magnetic North Pole in 1945. Wing Cdr. E. W. "Andy" Anderson, the senior navigator on the Lancaster, reported his magnetic compasses were working normally seventy miles from magnetic north's last reported position and still pointed northwest even when the *Aries* was directly over the spot.

"The Astronomer Royal had warned us that we might expect this; in fact by a complicated analysis he had placed the Pole about 300 or 400 miles further in this direction," Anderson wrote. "So we flew on for about an hour, the compasses gradually becoming more and more erratic."[29] Wing Commander McKinley later added, "When within some 200 miles of the charted position of the Magnetic Pole, all the compasses indicated the same so long as the aircraft remained steady, but the smallest acceleration or deceleration caused the free magnetic compasses to swing hard over and remain locked until steady flying was resumed."[30]

The flights of the *Aries* created a sensation. "McKinley was praised for attracting world admiration and gaining confidence in British aviation by his prowess and gallantry, materially contributing to the prestige of the RAF," a British newspaper recalled afterward.[31] The pilot remained in the RAF following the war and rose to the high rank of air vice marshal. Although no longer the bomber's pilot when the *Aries* shared the tarmac at Perth with the *Truculent Turtle*, Wing Commander McKinley in no small way helped pave the way north for American military fliers in 1946.

ELEVEN

JATO

THE *ARIES* LEFT GUILDFORD AIRPORT IN WESTERN AUSTRALIA at seven o'clock on Monday morning, September 23, with the afternoon *Perth Daily News* noting its departure on the front page. The famous Lancaster was bound for Melbourne and then home to England. The Americans remained. Commander Davies accompanied U.S. vice consul Rudolph Hefti later that day to call on Lord Mayor Joseph Totterdell. Navy technicians from the R5D meanwhile got busy overhauling the *Truculent Turtle*'s engines. If all went smoothly, they figured the plane might leave for Seattle on Wednesday or Thursday. Davies said reporters would learn the departure time at least twenty-four hours beforehand. Until then, he had scheduled test flights for Tuesday.

"Late this morning there will be a roar of two powerful motors, a cloud of dust, and the United States Navy's latest patrol plane, the Neptune, will speed down a Guildford runway on a test flight around the city and metropolitan area," the morning *Perth West Australian* told readers.[1] Actually the *Turtle* made two flights from Guildford that Tuesday. Davies was in the pilot's seat, with Gene Rankin as his copilot and Walt Reid and Roy Tabeling in the back. Vice Consul

Hefti and Captain Krieger of *Rehoboth* were aboard, too, although it's unclear whether for the first or second flight or for both.

Davies made his first takeoff under normal power, using only a quarter of the six-thousand-foot runway to get into the air without even a breath of wind to aid him. People saw the *Turtle* flying above Perth for an hour as the crew inspected the flight path and tested the plane's fuel consumption, engine performance, radio reception, and general airworthiness. Instead of landing back at Guildford, Davies set the Neptune down at Pearce aerodrome, an RAAF station twenty-six miles north of the city. He inspected the runway and the surrounding area, and welcomed on board Group Captain Hannah, the RAAF Western Area commander, who as Air Marshal Sir Colin Hannah would become governor of Queensland in 1972. The *Turtle* then took off again and returned to Guildford.

The *Turtle* had been fitted with four metal cylinders, sometimes called bottles, that were known as JATO units. Pronounced *Jay-tow*, the acronym stood for jet-assisted takeoff, which was a bit of a misnomer. The propellant-filled devices also and more accurately were called RATO (rocket-assisted takeoff) units. Hungarian immigrant Dr. Theodore von Kármán and colleagues at the California Institute of Technology at Pasadena began working on JATO in 1938. "Von Karman, with his sense of public relations, used the term 'jet' rather than 'rocket' because the latter was denigrated by many scientists as being 'Buck Rogers-ish,'" a modern account explains.[2]

Gen. Hap Arnold of the old army air corps was Kármán's friend and firm supporter. "Arnold pushed this program because it demonstrated potential for increasing the combat range of his heavy bombers," an air force historian writes.[3] Army Capt. Homer A. Boushey made the first experimental JATO takeoffs from March Field in Southern California in 1941. A year later, Kármán and his university colleagues founded a company called Aerojet to make JATO units. The General Tire & Rubber Company acquired Aerojet in 1945.

During the war in the North Atlantic, British forces made limited use of JATO units "attached to a Hawker Hurricane, mounted on some convoy freighters and launched at the sight of German anti-convoy bombers," according to an aviation historian. "The spent booster rocket fell to the sea, and the Hurricane could hopefully fly to land or ditch nearby so its pilot could be rescued by the convoy."[4] American forces used JATO, too, including on various flying boats involved in air-sea rescues.

"A pair of good-sized jatos are the equivalent of an extra engine at the crucial split second when a plane is leaving the water or the ground to become airborne," the *Saturday Evening Post* noted in 1945. The article recounted how a JATO-equipped Catalina had rescued eight downed airmen and their dog from a bobbing raft five hundred miles off the coast of Southern California. It also recounted how a twenty-eight-ton Martin Mariner flying boat stranded on a short Texas lake got airborne again thanks to JATO bottles attached to improvised brackets on the fuselage. JATO units, the magazine added, reduced the length of takeoff runs during testing "from 33 to 60 per cent, in the water, on landing strips and on carrier decks."[5]

Despite intense interest from the AAF and General Arnold, JATO was the navy's baby now; in fact, the *Turtle* had already tried it. "Several days before leaving [the] Lockheed plant in Burbank, California, we made a trial run using four JATO's to raise a gross weight of 70,000 pounds, which in itself [was] a record for twin-engine aircraft," Tabeling later recalled.[6] The navy contemplated a time it would operate new jet aircraft from its flattops, and the admiral commanding naval air forces in the Pacific expected that "long-range reconnaissance planes using jet-assisted takeoff features would make their appearance in the Pacific in the future."[7]

Navy technicians in Perth mounted a pair of white JATO bottles behind the wings on each side of the *Turtle*'s fuselage, slightly below and forward of the horizontal white bar and star of the national aircraft insignia. The two-hundred-pound bottles resembled small propane tanks. Each bottle provided the P2V with four thousand pounds of thrust—sixteen thousand extra pounds altogether—all delivered within twelve seconds during takeoff. The units would fire simultaneously when Davies pushed a jury-rigged button attached to his control yoke. "It's just like a small boy running as fast as he can," the pilot said, "and then suddenly a strong man picks him up by the seat of his pants and throws him into the air."[8]

A handful of aviation buffs stood watching near the takeoff point as the *Turtle* began moving down the Guildford runway late Tuesday afternoon. It was the first-ever JATO takeoff in Western Australia. Davies hit his JATO button as the Neptune reached seventy-fives miles an hour, with the propellant igniting with a tremendous *whoosh!* A description in the next morning's newspaper hardly began to cover the spectacle. "A tearing blast of thick, swirling, white smoke blanketed the runway in the rear as the plane hurtled forward with lightning acceleration," the *Perth Daily News* reported. "Rockets continued burning until the wheels were retracted. There was no provision for dousing them from within the plane."[9] The choking white smoke billowing behind enveloped the innocent onlookers and enraged a swarm of bees that sent everyone scurrying.

Watchers on the ground "contemplated what might happen if the aircraft hadn't succeeded getting off the ground because the duty runway pointed towards metropolitan Perth," a historian relates. Davies soon settled on using Pearce as the *Turtle*'s takeoff site for exactly this reason. Group Captain Hannah meanwhile considered the hop "the ride of his life."[10]

The test showed that JATO could boost a Neptune into the air after a very short run. A year and a half later, in April 1948 off the coast of Virginia, Davies would successfully launch another JATO-equipped P2V off the flight deck of the aircraft carrier USS *Coral Sea* with room to spare. A second Neptune followed, and both planes headed for a naval air station ashore. They didn't land back aboard the ship because even with adding a tailhook, setting down a P2V with a hundred-foot wingspan on a flattop's straight narrow deck was almost impossible. "That's the last time we ever try that!" exclaimed the only navy pilot who managed it.[11]

The inability to land land-based Neptunes aboard carriers would always be a problem. "The Navy had ordered a nuclear-capable attack aircraft, the North American AJ Savage, on 24 June 1946," writes naval historian Norman Polmar. "But it was not ready for delivery to fleet squadrons until fall 1949, and the Navy was unwilling to lose time in development of a carrier-based nuclear-strike capability."[12] So the admirals tried to claim instead that two carrier-based squadrons of modified Neptunes "represented a nuclear retaliatory force. In the event of atomic war they would take off, deliver their weapons, return to the carriers, and ditch in the water, where the crews would be picked up."[13]

THE JATO-LAUNCHED *TURTLE* CROSSED THE CITY AND FLEW out over the Indian Ocean, where Davies jettisoned the empty bottles as planned into the water. Secretary of the Navy James Forrestal later cabled him from Washington: "My congratulations on the successful completion of the first phase of your operation. On behalf of the Navy I extend to you and your fine crew best wishes for success in your long flight back."[14]

Back on the ground, Davies acknowledged the simple reason the navy had chosen Western Australia as the starting point for the historic long-distance attempt. "He said that it was about the most distant place away from the United States to suit the range of the plane," the *West Australian* reported. "Also, we were English-speaking people and as a result there would not be the same difficulties in preparing for the flight as in other countries."[15]

Davies also gave reporters a departure time for whenever the *Turtle* finally got away. The fliers would take off at sundown "in the last moments of daylight," so the Neptune could reach Milne Bay, New Guinea, north of Australia, in daylight the next morning. "This will reduce mountain and cloud menace in the New Guinea area," the pilot explained.[16] In addition, cooler evening air provided a bit more lift during takeoff and reduced turbulence later. The later hour also let the pilots rely on celestial navigation while flying across the continent at night. Sounding much like his AAF counterparts in Honolulu, Davies added that weather conditions had been favorable a week earlier but were deteriorating now, with contrary winds north of Midway Atoll in the central Pacific.

The *Turtle* didn't leave Perth as hoped the following day. After staying on the ground on Wednesday, it made the short hop to the RAAF airfield shortly before 5 p.m. on Thursday, September 26. "Its neat, streamlined appearance and high tail-plane were most noticeable when the Neptune made a quick 'pass' at the city at a speed above 300 miles an hour as it headed towards Pearce," the *West Australian* noted.[17] The R5D followed the P2V north to the Australian airbase. The *Turtle*'s crew hoped to take off for Seattle on Friday, "but we will wait until Saturday if reports indicate that the weather is improving further," one of them said.[18]

The American fliers and technicians made many friends during their brief time Down Under. Young Martin Laughton, however, would follow the *Turtle*'s flight home "with more eagerness than any other," according to the *Perth Daily News*. The eleven-year-old boy delivered newspapers to the team's Crawley headquarters. "They invited him to lunch in their mess, took him to Guildford to see their plane. While thousands of people looked at the Neptune from a distance, Martin was being shown over it by the men who will fly it."[19] Members of both aircrews signed the boy's autograph book, and the R5D men even took him up for a flight over Perth.

The newsboy couldn't accompany the aviators on the *Turtle*'s record-setting attempt, but another young Australian would be aboard. She was a nine-month-old, twenty-three-pound gray kangaroo from the South Perth Zoo. Juvenile kangaroos are called joeys, and this one, perhaps inevitably, was dubbed Joey, a male-sounding name that caused some confusion. Tabeling, for one, wouldn't realize that Joey was a female until after the flight, and Davies might not have known either. "Anyhow, it could be short for Joeline," a crew member remarked.[20] Commander Reid later called the little kangaroo "Miss Joey."[21] The fliers also contended afterward that the name might be the diminutive of Josephine, probably wishing that someone had named her Matilda in the first place.

The little animal and her crate added thirty-five pounds to the *Turtle*'s weight. "After all the trouble we went through to make the plane as light as possible, the crew decided to take that kangaroo," Lockheed's Robert Bailey said. The designer added later with a rueful laugh that the additional load was "where some of our gasoline went."[22]

But taking the kangaroo was a diplomatic rather than a personal decision. Found in the Narrogin district southeast of Perth, Joey was a goodwill gift from Western Australia to the United States. Her specially constructed crate was strapped into the *Turtle* before the final hop over to Pearce. The zoo's superintendent briefed the *Turtle*'s crew on caring for the kangaroo, which came with special grain food to keep her well nourished and comfortable during the long flight. Davies wasn't enamored, however, and learned the hard way how to avoid getting clawed by steering the feisty little marsupial safely from behind.

"To turn it to starboard you twist the tail to port. To turn it to port you twist it to starboard."[23] He added later that "I'd rather fly a plane any day than guide a kangaroo" and said Joey was "just plain vicious."[24] Davies also might have been tweaking the army's nose by taking the animal, because a kangaroo with boxing gloves was painted on the fuselage of General Doolittle's old B-29 *Challenger,* since flown by Colonel Irvine of the famous *Dreamboat.*

Joey was public relations gold for the navy and was mentioned in nearly every worldwide article about the flight. A box of champion Western Australian wildflowers was going aboard, too, adding yet more weight the *Turtle*'s crew couldn't have wanted. The fliers accepted the gift with grace and said they hoped the seventy varieties would last throughout the long flight, "so that they might present their wives and families in America with a bouquet of flowers picked only a few days earlier in a country thousands of miles away."[25]

Friday came and went without a takeoff. Unlike with the mechanically unlucky *Dreamboat* in Honolulu, the problem in Perth was strictly meteorological. "Weather conditions have again forced a postponement," Davies said after a long conference with navy weathermen on board *Rehoboth*. "It has been a touch-and-go decision."[26] The *Sydney Sun* on the opposite side of the country offered a bleak assessment: "Australian weather is still against 'operation turtle' and the Pacific situation is deteriorating."[27]

FLYING SUCH AN ENORMOUS DISTANCE OVER WATER COMplicated everything for Tom Davies and his small crew. Once past New Guinea and the Solomon Islands, the *Turtle* had to cross thousands of miles of trackless oceans and seas before reaching North America. The weather forecast for the *Turtle*'s flight, the navy said later, "was for a longer forecast period and covered a greater area than had ever before been undertaken for a non-stop flight."[28] The four-man crew was relying on radio stations at such isolated locales as Wake Island and Midway Atoll, and on weather ships designated "George," "Dog," and "1." These three ships, plus NAS Sand Point at Seattle, would receive the latest forecasts for transmittal to the Neptune when it came within range.

The *Turtle* had flown four long hops over several days to reach Perth: Burbank to NAS Barbers Point in Honolulu (where the Neptune was guarded overnight in a hangar to preserve the flight's secrecy); Hawaii to Majuro in the Marshall Islands; Majuro to Townsville, Queensland; and then the final hop to Western Australia. If the flight operation went according to plan now—and when did one ever?—once the *Turtle* was "wheels up" over Pearce, the fliers wouldn't touch down again until they reached North America or beyond. It was a daunting mission even for experienced naval aviators. The navy supported them by mounting a weather operation Down Under that matched the AAF's in Honolulu.

Rehoboth had brought a spare engine and propeller to Australia. The gear hadn't been needed, but the ship's compartments had since been transformed into a flight communications center that was even more vital. Cdr. Edwin T. Harding of Berkeley, California, arrived from the staff of the Western Sea Frontier to serve as the *Turtle*'s project aerologist. During the war, Harding had helped to establish the Navy Weather Central at Manus in the Admiralty Islands.

The navy's current weather capabilities were something of a hodgepodge. The weather service itself traced back to the 1840s and pioneering American hydrographer, oceanographer, and author Lt. Matthew F. Maury, who later served in the Confederate navy during the Civil War. "An organization called the Marine Meteorological Service actually existed in the 1800s," Harding later recalled. "From this, we have matured in the bureaucratic womb, in many queer and unusual shapes and forms, until, in July 1967, the Secretary of the Navy established the Naval Weather Service Command."[29] Harding himself later headed the Naval Weather Service as a captain after first commanding the navy's Hurricane Weather Central in Miami and writing a well-regarded book on heavy weather.

Harding collected weather data at Freemantle from sources in Australia and from the weather centers at San Francisco, Pearl Harbor, and Guam; the latter site radioed it to him aboard *Rehoboth*. Six navy radiomen handled the traffic. Commander Krieger, the seaplane tender's skipper, explained that the ship was a key link in a long chain connecting the *Turtle* to bases in the United States, "particularly for the reception of meteorological information covering the course to be flown by the plane. This information was being plotted daily, and the plot would be continued until the plane was well on her way to Seattle."[30]

The problem for the *Turtle*'s crew was that the weather picture Harding was assembling for them contained troublesome gaps. The weather bureau in Perth received good information via teletype on weather over Australia, but data on conditions in New Guinea and the Solomons was alarmingly slim. "The upper wind reports from this area were received spasmodically and from only one or two areas," the navy reported afterward. "There were no plane or ship reports available. Consequently the forecast for the leg from Cooktown, Australia to the Marshall Islands was based on inadequate data."[31] Harding was forced to rely as much on his own experience and intuition as on the insufficient data he was getting.

Rehoboth's sailors marked the data they did receive on enormous maps as if plotting a military campaign, which essentially they were. "It was revealed tonight that this will be the first time that a weather map on such a scale was charted and used," an Australian newspaper reported as September waned. "The weather is being calculated from reports coming in every hour from right across the hemisphere."[32] The *Perth Sunday Times* outlined the communications chain for readers. "Neptune will keep in radio contact with Australian aeradio stations on the first stages," the paper reported. "Hourly schedules will be kept with Guildford aeradio station, which will relay position reports to USS Rehoboth, at Fremantle. After passing Buna Bay [New Guinea], contact will be kept with the USN at Brisbane until completion of the flight."[33]

Like the AAF, the navy had drawn up rescue plans in case the *Turtle* was forced down by weather or mechanical problems. Officers said the Neptune never would be beyond the reach of air-sea rescue stations as the fliers crossed the vast stretches of sea between Perth and Seattle. But the plane first had to get into the air. Saturday passed without a takeoff, which was postponed after another long weather conference.

The *Perth Daily News* was reduced to reporting the assortment of food the manager of the local Dutch Club had prepared for the crew and packed into a picnic hamper filled with dry ice. The delicacies included crayfish with mayonnaise, two roast chickens, fresh eggs, "eight fillet steaks prepared in the Dutch manner and ready for cooking," a boiled ham, tomatoes, lettuce, bread, cheese, butter, olives from *Rehoboth,* two dozen apples, and an equal supply of oranges.[34] The paper added that the manager's wife had supplied the rationing coupons needed to purchase the meat and butter.

THE SUN ROSE EIGHTEEN HOURS LATER ON OAHU ON THE opposite side of the international date line. Bill Irvine nursed hopes at Hickam Field of finally getting away on the *Dreamboat,* now that the second ruptured fuel tank had returned from California and been reinstalled. The B-29's transpolar flight had been delayed so often and for so long that he would have to take off at dawn rather than at mid-morning as first planned. The change, Irvine explained, was "accounted for by the difference in the position of celestial bodies on which navigational fixes must be determined," the *Honolulu Star-Bulletin* reported. "Position of these bodies changes after September 15."[35]

Irvine said a test hop on Tuesday had shown the *Dreamboat* to be in "A-1 condition." But a crewman added that contrary winds along the long route meant "we couldn't have had enough gas left to taxi to our hangar after we reached Cairo."[36] So the crew stayed on the ground, waiting . . . waiting . . . waiting. Takeoff was pushed back to Friday; then it was delayed again. Saturday morning, while the *Turtle* sat on the tarmac at Pearce, conditions finally seemed to be right at Hickam Field.

"Radio and weather stations, ships at sea and listening posts half way round the world were notified. The plane was readied," the *Star-Bulletin* reported. "Four hours later Col. Irvine changed his mind. A low pressure area had suddenly moved into the path the Dreamboat will follow over Alaska. Sunday is another day."[37]

TWELVE

The Pacific

IT WAS THE BEGINNING OF SPRING IN THE SOUTHERN HEMI-sphere. An election occupied Saturday, September 28, along with sporting events ranging from horse racing to cricket. Sunday dawned warm and sunny, and the day was quiet as Sabbaths typically were Down Under. Winds were northwesterly near Perth; the only clouds in Western Australia were reported in the south at Cape Naturaliste and Cape Leeuwin. Temperatures in the city rose steadily from fifty-one degrees Fahrenheit in the morning to eighty-two degrees by two o'clock. Thermometers inched toward the century mark in northern parts of the state. At three o'clock, the barometric pressure in Perth was 29.989 inches (1015.5 millibars) and falling, indicating a high-pressure area that had dominated the region for several days was now weakening.

Tom Davies was tired of delays. "We figured ten days to ready the ship and to wait for good weather," he said later of the *Turtle,* "but the weather we wanted stretched over half the world."[1] The crew would have settled for any decent weather along their long route, and Commander Harding worked hard to find it. "On September 27th, their dedicated aerologist predicted that the weather would be satisfactory on the 28th, but better on the 29th with a slight tailwind component at the start," a short history of the flight says. "The 29th was chosen, and the ball was put in motion for a 6:00 PM departure."[2]

"With the rough draft of the forecast prepared, a lengthy discussion was held with the plane commander about the situation as forecast, emphasizing the effect on the flight of possible errors in analysis," a navy document adds. "Then the prognostic charts . . . were quickly drawn on board the Rehoboth and delivered to the Turtle just prior to takeoff."[3] Davies later praised Harding's performance. "Our meteorologist was perfect," he said. "His weather predictions were excellent."[4]

Like Guildford, the Pearce aerodrome had two six-thousand-foot runways plus a third of forty-five hundred feet. The *Truculent Turtle* sat parked in the Sunday afternoon sunshine at the end of one of the longer runways. Commander Reid later wrote that the Neptune was now "virtually a flying gas tank."[5] The patrol plane was designed with a "very large fuel capacity," Davies added. "For the record flight, however, a nose tank was installed in place of the six 22 mm. cannon, and extra fuel tanks were also installed in the fuselage in addition to tip tanks, giving a total capacity of more than 8,600 gal."[6]

The propellers were motionless as the ground crew clambered onto the wings to complete the fueling begun earlier. They had to finish the job on the runway because fully loaded (technically overloaded) with over twenty-five tons of fuel, the plane was far too heavy to taxi out without risking damage to the tires and landing-gear struts during a turn. The *Turtle*'s equipment included a dump system poking through the belly behind the bomb bay to jettison the gas in case of an emergency.

"Neptune took on petrol from three tankers from 1 p.m. to 4.20 p.m.," a Sydney newspaper reported. "She filled the nose, belly, tail, wings, and wing-tip tanks to the absolute limit."[7] Factoring in the fuel already in the lines, the *Turtle* carried over eighty-seven hundred gallons of hundred-octane gasoline and weighed more than eighty-five thousand pounds, nearly 60 percent of it being gasoline. This was the largest weight any twin-engine airplane had ever attempted to carry aloft, and it gave the *Turtle* an almost unimaginable wing load of over eighty-five pounds per square foot. *Science Illustrated* later noted that a well-known engineer had flatly declared not long before that a wing load of forty pounds was impossible. "By that he meant we could never get off the ground if the weight of the airplane and its load exceeded 40 pounds for every square foot of wing area," the magazine explained. "Actually that figure was topped some time ago, but conservatives then said we'd never get above 80 pounds."[8]

The four white JATO units attached to the sides of the *Turtle*'s blue fuselage were also all filled and ready to go. The bottles weren't the only items the pilots would jettison during the flight. The plane's two cylindrical wingtip fuel tanks were expendable too. Designed for Lockheed's twin-tailed P-38 Lightning fighter, the streamlined tanks were slung from bomb shackles attached to the very end of the Neptune's long wings. The tanks extended the *Turtle*'s range with almost no loss in airspeed. Although built to hold three hundred gallons of gas each, due to structural limitations, they would carry only two hundred gallons apiece during the flight.

The tip tanks reminded some onlookers of bombs or torpedoes and others of floats affixed to Catalinas and other flying boats. Davies had explained that the tanks were made of nylon—"much to the ladies' disgust," he had added. Since the synthetic material was still scarce following the war, few sheer stockings were yet available. The tanks were exceptionally light and strong, Davies said, but basically were "just bags of juice."[9] Because of weight distribution and wing load, the *Turtle* performed best if the tanks remained full early in the flight. The pilots would use fuel from the fuselage tanks before tapping the wingtip tanks; then they would drop them once the tanks were empty. Lockheed built later Neptunes with permanent wingtip tanks and a distinctive tail boom to house electronic gear, creating a profile that was familiar to a generation of navy fliers.

Shortly before takeoff time, somebody (likely one of the ground crew) discovered that the Neptune's directional gyro wasn't working. Consequently, "the R.A.A.F. lent one to the crew of the 'Turtle,'" a British aviation publication reported. "It is probable that this was a Sperry, although it might have been a Reid and Sigrist instrument. For a flight of this kind the Directional Gyro is one of the most important pieces of equipment in the aeroplane."[10]

Loading kangaroo Joey was somewhat troublesome too. First, the animal showed "a disposition to avoid the trip and go into the bush," the *New York Times* reported.[11] Then a "furiously barking" dog objected to the little marsupial's presence at the RAAF aerodrome.[12] A man in a pith helmet held onto the shaggy canine while a zoo official kept a firm grip on Joey's tail, thereby avoiding bloodshed and poor international publicity. Joey's crate was safely stowed in the *Turtle*'s tail.

The Western Australian wildflowers went aboard too. The crew, however, was traveling light. Tabeling's luggage consisted partly of a paper grocery bag containing his orders, passport, and a change of clothes. The *Turtle*'s crew also took aboard a batch of letters bearing an Australian stamp and a Perth postmark intended for VIPs, family, and friends in the States. "In the letter they explained the purpose of the flight. The cover pictured a plane of the 'Truculent Turtle' type with the inscription 'Navy P2V Special Air Mail,'" a newspaper reported afterward.[13]

Commander Davies also carried a letter to President Truman from Lord Mayor Totterdell expressing friendship from the people of Perth and Western Australia. No other city in Australia, the letter said, had "greater cause to remember the wartime help of America, for it was almost from the very portals of this city that the courageous officers and men of the submarine flotillas and air squadrons of the United States made such devastating sorties against our common enemies in the Indian Ocean."[14] Davies carried a second letter, too, from the lord mayor to the mayor of the *Turtle*'s presumed destination of Seattle.

Lieutenant Horner, *Rehoboth*'s doctor, wandered about the runway, filming in color with his personal movie camera. He captured the *Turtle* gleaming in the sunshine while parked at the very back edge of the runway with a low forested hill beyond it and a yellow refueling truck parked behind the starboard wing. A milling crowd of civilians, including a young boy in short pants and Aussie and American service members, inspecting the plane. *Rehoboth*'s Captain Krieger chatting to two young women wearing bright dresses. Joey being bodily lifted into the *Turtle*. An RAAF sentry armed with a rifle and bayonet gesturing to civilians. And the four khaki-clad fliers posing beneath the Neptune's wing: Davies in a billed cap standing beside Rankin, with their arms around each other's shoulders and flanked by smiling Reid and Tabeling. Several press photographers prepared to capture the takeoff as well. Lieutenant Horner came to believe his footage (particularly of the jet-assisted takeoff) was unique, since the official film that reached the United States was already on board the plane, having been shot during the JATO test at Guildford.

Before the takeoff, the lord mayor sealed a barograph to be carried on the *Turtle*. This bit of gear for recording altitude and temperature in the air was the usual method for certifying a plane's long-distance flight. "Well, the next time we touch earth, it's going to be America," Davies said as he climbed into the plane.[15] He settled into the pilot's high seat on the left side of the cockpit. Rankin took the copilot's seat opposite him. The other two pilots sat in the back, with Reid manning the fuel-dumping station near the aft fuel tanks. He knew he stood little chance of survival if the *Turtle* crashed before it got airborne.

By six o'clock, the sun was setting quickly, and the wind had died to a negligible seven knots. Anxiety and expectations were high, although as the *Perth Daily News* reminded readers, the flight "had not been heralded by the Navy as a try for the world long-distance record."[16] The crowd quietly dispersed as Davies started his engines. Technicians listened keenly for any problem as the two Wright Cyclone 18s warmed up. The ground crew pulled the chocks out from under the Neptune's wheels when Davies gave a thumbs-up signal from the cockpit. Rankin sat for a moment, thinking about his wife and three young daughters at home. When he'd asked the eldest, not yet three, for something special of hers to bring along for luck, the little girl had fetched her favorite pair of underpants, which were now tucked away in his flight bag. "What am I doing here in this situation?" Rankin wondered. The tower cleared the *Turtle* for takeoff and wished the crew a cheerful "good luck."[17]

The sun was nearly down at ten minutes past the hour. The P2V was designed to take off in less than a thousand feet but not while carrying such a massive weight. While the navy thought the plane might get airborne within thirty-five hundred feet, the pilots would let the *Turtle* determine its own liftoff point. The RAAF runway pointed northeast to southwest toward the Indian Ocean. The pilot who had built model airplanes as a student back at Cleveland's East High School now tromped on the brake pedals of a powerful aircraft. Davies pushed his throttles forward and waited until his engines reached full power before releasing the brakes. The *Turtle* trundled rather than leaped forward.

Rankin called out speed and distance as they passed the thousand-foot runway markers, checking their progress against notes he'd jotted on a clipboard. The *Turtle* was committed beyond the second marker, because Davies could no longer stop without running out of runway. It was crash or fly.

The *Turtle* passed the third marker doing about eighty-five knots, or nearly a hundred miles per hour. Davies punched the JATO button, unleashing twelve seconds of roaring power. Spectators across the aerodrome held their breath as the *Turtle* stayed on the ground. The Neptune picked up speed, and at forty-five hundred feet, the fliers felt the weight lighten on its wheels. "She is supposed to fly at 115 knots," Davies said. "Pull up the wheels."[18] Rankin hit the actuator. "If he remembered to do so in all the excitement," a former navy pilot and author writes, "CDR Davies likely tapped the brakes to stop the wheels from spinning before the wheel-well doors closed, just as the JATO bottles burned out."[19]

Horner's home movie, shot from a long distance, showed a thick low-lying cloud of white exhaust trailing the plane. Davies said later he believed only two or three of the four JATO bottles actually fired, but if so, it was enough. "Failure of one or two units would not cause any difficulty in clearing the field, nor would a single failure cause swerving on the take-off," Rankin said later. "To say the least, the take off, to those watching it, was spectacular."[20] Tabeling decades later called it "a very short takeoff for an overweight plane."[21]

The P2V lifted a little way into the air and roared toward the sunset. It lingered above a low stand of trees beyond the airfield for what seemed an agonizingly long time. "Boy, was I glad to see those wheels leave the ground," a navy technician said. "Believe me, I was worried for a minute."[22] A decade later, Lockheed executive and former Vega general manager Courtlandt Gross said, "It's one of the ageless wonders of aviation how that plane ever got off the ground."[23] Nobody was happier than the *Turtle*'s crew.

"We all heaved a sigh of relief when that airdrome disappeared behind us and we were steadily gaining altitude," Tabeling wrote.[24] To Davies, the takeoff was the biggest thrill of the trip. "After we were airborne," he said, "the record distance was assured since we had gasoline enough to fly at least to Bermuda."[25] But they didn't have as much as they expected, and Davies realized almost at once that somehow two hundred gallons of gas hadn't been pumped into the Neptune. From a short cryptic remark he made later, it's unclear whether the mistake involved a wingtip tank, a wing fuel cell, or a temporary fuel tank stowed inside the fuselage. But two hundred gallons was a minuscule portion of the enormous fuel load and, at the time, seemed insignificant.

The *Turtle* slowly gained speed and altitude while flying eighteen miles to the coast. Davies jettisoned the four JATO bottles once over the water and made a shallow left turn onto a northeasterly heading over Perth. Davies climbed to nine thousand feet and set the automatic pilot, a device the pilots had nicknamed "George." They would cross the forbidding outback and northern Australia in the dark, navigating over the continent by using radio beacons at Alice

Springs, Northern Territory, and Cooktown, Queensland. According to navy aerographers, the pilots had "little hope for helping winds through the strong and persistent Southeast and Northeast Trades" while the *Turtle* was between thirty degrees south and thirty degrees north, the latitudes running parallel to the equator from north of Perth to south of San Diego.[26]

In Hawaii on the other side of the international date line, the *Dreamboat*'s commander was naturally disappointed the AAF hadn't gotten away first. But Bill Irvine was a good sport when he learned of the *Turtle*'s takeoff. "I wish those boys all the success in the world," he said. "I'm sorry that we couldn't get off at the same time to make things more interesting."[27]

NAVY PLANNERS HAD CONSIDERED USING SINGLE-ENGINE flight during much of the record attempt to extend the Neptune's range but dropped the idea as too risky. The *Turtle* pilots eased off the throttles to conserve fuel and kept their average speed relative to the ground or the sea at 200 to 220 miles per hour. The four aviators saw few or no lights below as they flew over vast and arid Australia. They would have been almost as alone crashing in the outback as they would have been ditching at sea.

An hour and a half after takeoff, the Neptune passed sixty miles north of tiny Kalgoorlie, Western Australia, at six thousand feet. "Naturally there wasn't much sleep the first night," Tabeling recalled, "because we were all too excited to do anything except maintain constant check on our instruments and fuel."[28] Late that evening, the plane passed near an unnamed point in the desert where Western Australia, South Australia, and the Northern Territory all meet. Soon after midnight, they passed Alice Springs, a station so small and remote in the heart of Australia that on the hop down the crew hadn't seen it even in daylight. Trying now to make contact by radio, Davies "discovered Alice Springs had no voice transmission. Contact from the station to the Turtle was by international code."[29]

Farther on at Queensland, the York Peninsula pointed like a dagger toward the belly of the divided island today called Papua and Papua New Guinea. The *Turtle* crossed the peninsula's southeastern coast ten hours and five minutes after takeoff. The P2V winged out over the Coral Sea at daybreak thirty miles south of Cooktown, named for the British sailor and explorer Capt. James Cook. The crew was now more than twenty-one hundred miles from Pearce. Before switching from Australian frequencies, Davies radioed his appreciation for the cooperation various stations had extended to his crew as they had crossed the continent.

The Neptune began crossing the ghostly wakes of key naval engagements during the recent war. Far off the starboard wing sprawled the watery site of the Battle of the Coral Sea, the U.S. Navy's first big fleet action against Japan and the world's first major air-sea battle. In early May 1942, having already established a foothold in northern New Guinea, Japanese forces moved toward Port Moresby on the island's southern coast, considering it a stepping-stone for an invasion of northern Australia. The naval battle by U.S. and Australian forces to stop them ended in a stalemate.

The opposing task forces never caught sight of one another. The U.S. force lost the flattop USS *Lexington* and sustained damage to the *Yorktown*, but it crippled the Japanese carriers *Shōkaku* and *Zuikaku*, participants in the December 1941 attack on Pearl Harbor. "On the Japanese part they managed to sink more American ships than they lost," an Australian naval history says, "whilst the Allies not only prevented the Japanese from achieving their objective, the occupation of Port Moresby, but also reduced the forces available to the Japanese for the forthcoming Midway operations."[30]

The *Turtle* crossed the narrow northwestern section of the Coral Sea amid tall, puffy, white cumulus clouds. The pilots dropped the empty wingtip fuel tanks, which fell from the shackles like dummy bombs down into the sea. "Everything going fine," the crew radioed during the morning. "Kangaroo riding comfortably."[31] The P2V reported passing over Milne Bay at 8:45 a.m. local time, with the far southeastern tip of mountainous New Guinea now off the port wing.

The plane continued through what the navy called "typical tropical conditions" in the New Guinea–Solomon Islands area.[32]

Newspapers in America and Australia tracking the *Turtle* speculated whether Seattle really was its final destination. "Navy planes will be sent up to rendezvous with the ship as it nears the West Coast," the *Honolulu Advertiser* reported. "Some idea of the eventual landing point may be learned then when the physical condition of the crew and the gasoline supply is determined."[33]

Beyond New Guinea, the *Turtle* crossed over the Solomon Sea amid squalls and headed toward green jungle islands that had seen terrible combat between the Japanese and Allied forces during the war. The flight grew rougher around noon as the P2V skirted a thunderstorm and flew past the northern tip of Bougainville. "The weather around Bougainville was just as tough for us as it was for the boys during the war," Davies said later.[34] The *Turtle* punched through the storms and droned steadily onward.

Over six thousand miles of deep water lay between Bougainville and Seattle—across the South Pacific Ocean, the equator, and the northern Pacific. The Neptune was a fine aircraft for making so long a trip. It was equipped with a small electric stove, bunks with foam-rubber mattresses that helped cancel vibrations, and a chemical toilet aft. The crew ate well once they had the chance. Someone had given the fliers a box of lobster meat before takeoff; they ate it first before the meat spoiled. "We had all the grilled steaks we wanted, ate three good hot meals a day, drank hot coffee and fruit juices and even shaved and washed—everything a man wants in his own home," Davies said.[35] The fresh food lasted about thirty hours, after which the crew mostly ate canned soup. But Joey the kangaroo grew unhappy, flopped down in her crate, and refused to eat.

The crew stood a navy watch rotation of four hours on and four hours off. "Pilots rode ahead of noisy propellers, and the cabin was insulated on a par with modern transports. . . . Instead of head sets, apt to get heavy with the hours, they wore light plastic ear plugs wired for sound with a cut-out to eliminate loud noises," W. H. Shippen

Jr. wrote for the *Washington Star*.[36] During the daytime, the off-duty pilots "sat around and talked and did various other things to pass the time," Davies said.[37] At nighttime they slept. The crew had been issued Benzedrine pills to keep alert but never took them.

The pilots kept the plane properly balanced by manipulating pumps and valves to shift gasoline between the various temporary and permanent fuel tanks. George, the automatic pilot, meanwhile handled 90 percent of the flying. Once past the Solomons, the navy reported, the *Turtle* entered a blackout area, where it was "unable to contact any of the very few weather reporting and navigational radio stations."[38] Davies and his three companions were on their own during their lonesome passage across the Pacific. "Our radio equipment was not powerful enough to report our position to Midway or Honolulu; consequently, we were in a radio blackout for about 18 hours," Rankin wrote later.[39]

The Neptune crossed the equator as it approached the Marshall Islands, which American forces had wrested from the Japanese in early 1944. Davies's hope of seeing clear weather ahead was dashed. "When they headed out across the South Pacific, beyond the Solomons, Commander Davies continued, the Turtle ran into an equatorial front which was 'about 250 miles of bottled up cumulus,'" the *West Australian* reported.[40] Reid later recalled that while passing through this buffeting, "we began to worry about the plane's ability to take such a pounding with so great a load. We were afraid the fuel tanks might leak, but nothing happened."[41]

The *Turtle* flew through the Marshalls at about eight thousand feet, passed between the atolls of Majuro and Kwajalein without spotting either, and winged into a second nightfall. The plane radioed that "the weather was good and the plane's performance excellent," the *West Australian* informed readers in Perth. The paper added that if "misfortune does not intervene," the issue no longer was whether the *Turtle* would break the *Dreamboat*'s Guam-to-Washington nonstop long-distance record "but how far it will go past that distance before shortage of fuel or pilot fatigue necessitates a landing."[42]

The Neptune crossed the dateline shortly before midnight, jumping back twenty-three hours and beginning September 30 over again. The weather remained poor until the plane passed over French Frigate Shoals in the Northwestern Hawaiian Islands southeast of Midway Atoll. (Newspaper reports placed the *Turtle* 550 miles from Midway, but the shoals are 760 miles away.) French Frigate Shoals had a small airstrip that was used during the war as a refueling stop on the way to Midway, which was the scene of another pivotal air battle between American and Japanese carriers that never sighted each other. The flight was a history lesson the *Turtle*'s veteran crew hardly required. "While the world slept, went to work, and slept again," an Australian newspaper noted, "the U.S. Navy plane, 'Truculent Turtle,' has covered 5300 miles on its 9000-mile nonstop flight from Perth to Seattle (Washington State)."[43]

The *Turtle* had used enough gas now to stay afloat if forced by a mechanical failure to ditch in the calm seas below. Appropriately, Midway Atoll was indeed about halfway between Perth and Davies's destination of Bermuda, 650 miles off the North Carolina coast. The *Perth Daily News* was embargoed until the *Turtle* reached the atoll, but now it could reveal that Davies's real goal lay out in the Atlantic Ocean. "It's tough to Midway," the commander had told an aviation writer back in Australia. "From then on it's a lead pipe cinch."[44]

The fliers abandoned their experiment with pressure pattern navigation after the weather forced them too high. "It was found impracticable to use this system except for one very short interval," *Aviation News* reported. "Following a single heading depended upon the radio altimeter which did not function above 8,000 ft. Most of the flight was between 8,000 and 12,000 ft." The *Turtle* didn't pick up the expected tailwinds either but instead "bucked headwinds a considerable part of the time."[45] Joey began to perk up in her crate once the plane reached calmer weather near Hawaii. "Well, the damned kangaroo has started to eat and drink again," Reid reported around noon. "I guess she thinks we're going to make it."[46] Tabeling seemed to like the little animal better than the other three and considered her "comic relief."[47] He reported that she eventually ate out of their hands, but probably it was only from his.

About this time, the *Turtle* had a major setback, albeit one apparent only in hindsight. It was a systems failure rather than a mechanical problem. Emerging from their long communications blackout after leaving Australia, the crew radioed the Civil Aeronautics Administration (CAA, the forerunner of the modern Federal Aviation Administration) at Pearl Harbor for a weather update. "The administrative secrecy which surrounded the initial stages of planning of the flight left this organization not fully alerted and they contacted the Weather Bureau at John Rodgers Airport, instead of the Navy," a navy report explains.

The duty forecaster replied with a hurried rundown based on the current weather map. The briefing amounted to this: "Icing conditions over Seattle and the Cascades. Continue on or north of the great circle course to Seattle. Strong NW winds will be encountered before reaching the Coast, where they will back to Southerly to Southwest." The information was so short and vague that the crew concentrated on the icing issue, a danger that could imperil the plane. Because of distance and "generally poor reception," according to the navy, the *Turtle* didn't contact the weather ships up ahead for the latest forecast, which was prepared by NAS Alameda across the bay from San Francisco.

Davies reacted cautiously based on the limited data and decided to veer from the great circle route they had followed to this point. He turned toward the Oregon-California state line, about 450 miles south of Seattle, rather than continuing straight ahead. The change took them out of radio range of the weather ship George, which they had planned to overfly 1,700 miles off the West Coast. According to the navy's later analysis, the new course meant that "advantage was not taken of tail winds on the North side of the Pacific High."[48]

The deviation also meant the *Turtle* had to fly farther to reach either Washington or Bermuda and with less of a push from Mother Nature. The Neptune dipped its wing into a starboard turn toward the long North American coastline still some three thousand miles away.

THIRTEEN

Landfall

THE *TRUCULENT TURTLE* FLEW HOMEWARD THROUGH another long day, its two engines "scarcely missing a beat."[1] The flight over the North Pacific Ocean was so routine and uneventful the pilots barely mentioned it later. They switched from visual to instrument flying about two hundred miles west of North America. The Neptune neared Northern California after nightfall on Monday, September 30, after its second day aloft. The West Coast was on Pacific standard time, sixteen hours behind Perth, having switched from daylight savings time on Sunday.

Approaching the coast, the *Turtle* slowed in a rainstorm that would soak the Bay Area early the next day. The navy hadn't heard from the plane since the crew radioed the CAA in Hawaii nearly fourteen hours earlier, and the Neptune missed an expected afternoon check with the weather ship George. As anxiety over the silence mounted, the commander in chief of the U.S. Pacific Fleet issued an order at nine o'clock to all his western radio stations: "Make every effort to contact and advise."[2]

Paine Field in Washington picked up a faint signal only nine minutes later. The *Turtle* radioed that it had swung south off its course over the Pacific to avoid storms and was now about seventy-five miles off the California coast, north of San Francisco. Half a dozen other stations soon picked up the plane too. "We are climbing to 12,000 feet," the crew reported. "Have you any reports of icing? I am unequipped with deicing equipment."[3] The Neptune called again at 9:11 p.m.: "Hello, radio Oakland, this is Navy 9082, arriving west leg Red Bluff range, using northwest leg of Oakland Range for check, reporting to United States from Perth, Australia."

A station at Bainbridge Island on Puget Sound heard a partial transmission at 9:13: "Will check in Red Bluff range after checking weather across United States. Our fuel supply and other—."[4] The navy expressed relief and some surprise at Davies's reported position. "The plane was somewhat south of the route which Navy officials believed it was following and approached the coast about 125 miles north of Sacramento, coming in on the Red Bluff radio navigational range."[5]

The Neptune made landfall about 9:30 at Point Cabrillo above Mendocino, although the pilots never saw the darkened coastline below. At 9:45 the navy announced that the P2V had broken the *Pacusan Dreamboat*'s long-distance nonstop record set the previous November. "When it checked in with radio stations north of Sacramento the big two-engined plane had covered 9,200 miles in 43 hours since its takeoff near Perth," the *Washington Star* reported.[6] The navy added, however, that it would claim no record until the Neptune landed safely—whenever and wherever that might be. With the plane still in the air, the *West Australian* back in Perth declared the flight a "magnificent achievement . . . [and] each successive hour the Neptune flies beyond its initial goal [of Seattle] will add still more to aeronautical knowledge and luster to the crew's achievement."[7]

Roy Tabeling manned the radio as the *Turtle* headed across the darkened continent. He contacted a ground controller at Williams, California, seventy miles south of Red Bluff, and requested instrument clearance to continue eastward. "This is the first airplane ever to fly nonstop from Australia to the United States," the United Press reported him saying.[8] But Tabeling had a tough time convincing the controller his plane really had flown nonstop all the way from Western Australia. He had to tell her three times that they had started the flight in Perth. The controller exclaimed in disbelief that the city was halfway around the world. "No. Only about a third," the lieutenant commander told her.[9] His answer still wouldn't do. "She just wouldn't believe us," Gene Rankin recalled. "We finally had to tell her to get in touch with the Navy in Alameda."[10]

The *Turtle* encountered another systems failure over California, one that again was apparent only in hindsight. This second failure involved an airway designated Green 3, the main air corridor between San Francisco and New York City. The crew requested information on weather between Reno and Salt Lake City plus clearance through the Sierra Nevada via the Donner Pass. The difficulty was that the naval radio station at Alameda "was too limited in power to reach the plane and to direct them over the more favorable course via Boise and North Platte," a navy analysis found later.

> Air Traffic Control (again, insufficiently briefed on the flight) was more concerned with getting the plane cleared and in reporting the Green 3 Airways weather than in finding and giving them the best course for the flight. Clearance to Donner Summit was denied, whereupon the plane went "off airways" to Donner, where clearance was then granted to Reno. Had the Aerological Officer at NAS, Alameda, been able to talk directly to the plane about the Boise route, the flight would have avoided some of the turbulence and icing encountered on the Green 3 Airways, as well as saving the loss in miles caused by turning southward off course and into the adverse winds between Red Bluff, California, and Reno.[11]

Directing a flight change toward Idaho would have added at least two hundred miles to the flight, the navy estimated, but also would have given the *Turtle* better flying conditions, with flight-level winds generally blowing from the northwest or north-northwest from Boise to Washington DC. On the silver anniversary of the flight, Davies would recall that the *Turtle* "had to be flown in both hemispheres, with their offsetting prevailing winds," and added, "I figure we netted about a five knot head wind on our flight."[12]

The *Turtle* passed north of Williams at 10:38 p.m. Pacific standard time. The crew aimed for the Donner Pass in the mountains northwest of Lake Tahoe, "seeking to ride the beam through that opening of the high Sierras."[13] Here an ill-fated party of immigrants led by George Donner and Jacob Donner had become trapped during the heavy snowfalls of 1846–47, with the survivors reportedly resorting to cannibalism to survive. The *Turtle*'s crew at least had a kangaroo to eat if forced down and stranded. The *Turtle* passed uneventfully through the Sierra and continued eastward.

Having swung south from the original great circle route, Davies now flew along a shorter great circle route from Donner Pass eastward to Washington DC. If he continued beyond the final 2,300 miles to the capital, he would need to turn somewhat southward again to reach Bermuda. The navy told journalists the weather ahead looked good: "An 8-mile-an-hour tail wind, with scattered clouds at 10,000 feet, and 30-mile visibility, with a high pressure area east of the mountains virtually to the Atlantic seaboard."[14] The Neptune's progress would make news across the States come morning. "The Truculent Turtle was reported over Humboldt, Nev., about 12:35 a.m. by the Elko (Nev.) C.A.A. tower," blared a bulletin beneath a banner headline on the front page of the *Los Angeles Times*. "Within a few minutes the Navy plane was out of the Humboldt range and expected to contact Elko shortly."[15]

The crew was flying at twelve thousand feet and approaching Ogden, Utah, on the eastern shore of Great Salt Lake, when suddenly the night turned remarkable even by *Turtle* standards. The bulletproof windshield began glowing with what looked like greenish-blue flames. For a terrible moment, the crew thought the plane was ablaze; then they quickly realized it was harmless Saint Elmo's fire. Known to mariners for centuries and to aviators in recent times, it's a natural phenomenon sometimes encountered in thunderstorms when electrical voltage acts on a gas.

Sailors see Saint Elmo's fire flickering at the tips of masts and spars, aviators from their propellers or wingtips. Radio receivers and the air hiss and crackle with static. The spectacle is popularly connected with the Greek myth of twins Castor and Pollux and their sister Helen of Troy—all three being patron deities of sailors. Its name, however, derives from an Italian mispronunciation of Saint Erasmus, the patron saint of sailors in the Mediterranean. The spirit Ariel speaks of the eerie phenomenon in Shakespeare's *The Tempest*:

> I flamed amazement; sometime I'ld divide
> And burn in many places; on the topmast,
> The yards and bowsprit, would I flame distinctly,
> Then meet and join.[16]

When Saint Elmo's lights appear, a navy magazine explained, "sailors take it as a sign that no harm can befall them and they will come safely through the storm, guided by their patron saint."[17]

The Neptune's windshield in effect was a large illuminated neon sign. Later inspection revealed no damage to any part of the *Turtle*. Davies told friends later that when the Saint Elmo's fire lit up the cockpit, the crew's hair literally stood on end. "Of course, I knew that we were safe," he said, "because the gas tanks were well protected inside the plane, but it's hard to be objective when you are up in the clouds at night in a cockpit and something like that hits you."[18]

IN THE WEE HOURS OF THE NIGHT, EASTERN TIME, THE fliers' wives were unaware of the fiery encounter. Just as astronaut spouses would do a generation later, they had gathered Monday in hopes of greeting their husbands when they touched down sometime on Tuesday at NAS Anacostia in the southeastern part of the District of Columbia. The women were a trio rather than a quartet because Roy Tabeling was unmarried. "Although no designation other than 'the United States' has been announced by the Navy, there were reasons to believe that Washington is the hoped-for nonstop goal in view of the fact that wives of three of the four crew members have assembled here," the *Washington Star* reported earlier in the evening. By now the women might have heard that the *Turtle* was aiming for Bermuda.[19] Following the norm of that era, nearly all the newspapers identified the wives simply as "Mrs." rather than adding their first names.

Eloise Davies was the wife of the *Turtle*'s command pilot, so Virginia Rankin and Ruth Reid gathered by custom at her home in suburban Chevy Chase, Maryland. Eloise was a "navy brat" who understood the duties and responsibilities of a commanding officer's wife. Her late father was Rear Adm. Robert H. English, a member of the U.S. Naval Academy's class of 1911 who later commanded the light cruiser USS *Helena* during the Japanese attack on Pearl Harbor. He kept his torpedoed ship afloat despite swarms of Zeros, Vals, and Kates. Promoted to admiral and command of the Pacific submarine fleet, English died in an air crash thirteen months later during a violent storm over Northern California. His daughter immediately joined the navy's new Women Accepted for Volunteer Emergency Service (WAVES) and first wore her uniform during his funeral at Arlington National Cemetery.

Ensign English was an alumna of the prestigious Punahou School in Honolulu and Sweet Briar College in Virginia. The navy assigned her to public relations duties with the photo lab in Washington. She was working at the White House when illustrator Norman Rockwell visited late in 1943 to document a typical day at the mansion for the *Saturday Evening Post*. The acclaimed artist painted colorful little vignettes of senators, military officers, Secret Service agents, journalists, a visiting beauty queen, various others, . . . and Eloise, sitting primly in a red leather chair and wearing crisp navy whites. "What has Miss America got that Wave Ensign Eloise English hasn't got?" the caption asked, earning the young WAVE no end of teasing.[20]

Tom Davies read the magazine at his quarters in Brazil and vowed to look up the beautiful ensign once he returned stateside. He married her in 1945, and the bride used her father's sword to cut their wedding cake. Now Eloise was a busy navy wife and mother of a six-month-old son, Tommy Jr. Later she would use the G.I. Bill to study law and begin her career at the U.S. Department of Justice.

Commander Reid's wife, the former Ruth Critchfield, was equally impressive. She was a graduate of Eastern High School in Washington and was later a Phi Beta Kappa and voted the most outstanding woman of 1936 at George Washington University. Ruth earned a master's degree in physical education from Scripps College in California and taught physical education before the war at Redlands University. Senator Shipstead of Minnesota, her husband's uncle, was the best man at their wedding in July 1939. Five years later, Ruth accepted an appointment as the director of girls' and women's recreational activities for the Girls' Club in suburban Alexandria, Virginia. She would go on to work many years as a field director for the Girl Scouts. Ruth wouldn't say what she thought about the Perth flight, only that she hoped that Walt "just gets back."[21]

Commander Rankin's wife, the former Virginia Watson, was "a Southern belle working as an executive secretary in Miami" when the couple met and married shortly before the war. Virginia too was now a navy wife and mother but frankly admitted she hadn't wanted Gene to make the flight, which she considered dangerous. The Rankins would raise five daughters in all, and Virginia said three decades later that they "kept me rather busy until just recently."[22] She, Eloise, and Ruth gathered around a globe at the Davies's house to track their husbands' world-straddling flight on the *Turtle,* as the Navy Department kept them informed of the plane's latest known position.

By the time their husbands saw the Saint Elmo's fire, Virginia Rankin and Ruth Reid probably had returned to their homes for much-needed sleep. Eloise Davies meanwhile kept in touch by long-distance phone with Tom's parents, David and Katherine, who sat huddled around a radio in Cleveland. Perhaps Gene Rankin's parents, John and Agatha, were still awake too in Oklahoma. The couple anxiously traced the *Turtle*'s progress, with their "ears glued to their radio to catch all the news broadcasts and bulletins . . . following as closely as possible the flight of the two-engined navy bomber."[23] During the war, they had enjoyed the dubious benefits of time and distance. But the Rankins would learn very quickly if anything happened to their son's plane tonight.

THE *TURTLE*'S PILOTS FLEW ON INSTRUMENTS THROUGH exactly the sort of conditions Davies had feared and tried to avoid by turning off his original great circle route over the Pacific. Ice built up on the wings and fuselage as the *Turtle* flew through blinding snow and freezing rain. "We had been told there was no ice in the weather or the flight might have ended earlier," the commander said. "When we got in it, it was too late. So we just kept going."[24]

Pilots fear ice with good reason. It decreases lift and thrust while increasing drag and weight. A plane that accumulates too much ice literally falls from the sky. During the war, the AAF and the General Electric Research Lab in Schenectady, New York, spent a great deal of time and money studying the problem of aircraft icing (especially troublesome in the Aleutians) as well as cloud formations, icing nuclei, cloud physics, and more. An observatory atop Mount Washington, New Hampshire—location of the fiercest weather in the continental United States—was an extremely useful test site. Thanks to General Electric's efforts and much additional icing research since the war, all commercial aircraft and many private planes today are equipped with expandable boots or other tools for shedding ice. But the *Turtle*'s deicing gear had all been removed to save weight and extend its range.

The *Turtle*'s crew kept a sharp watch on their flight dynamics as they bucked over Utah toward the Rockies. Near Ogden, the port engine's tachometer abruptly went haywire. Davies aimed a flashlight out his window and saw the prop whirling normally. "Just as I was about to feather the left propeller I realized that the engine was running all right and that it was just the tachometer that was out," he said later. "Shortly thereafter the right engine quit—for about 30 seconds—from ice."[25] Davies dropped the plane's speed and switched to an alternative heating system, which brought the engine back to life. The incident was nonetheless unsettling. Rankin later wrote that they were forced to boost power 80 percent simply to stay in the air, and he estimated that the *Turtle*'s heavy fuel consumption during the three hours of turbulence had cost them five hundred miles' flying distance. Davies observed that the *Turtle* "had all the weather we could find, including 1,000 pounds of ice."[26]

"We accumulated so much ice over the Sierras and the Rockies that we could not see our way around at all," Tabeling recalled years later, "but we just kept going and broke out on the other side after many hours."[27] The Neptune finally crossed into southern Wyoming at 2:30 in the morning, mountain standard time, flying above ten thousand feet with clearer weather ahead. At 3:30 a.m., the *Turtle* was five miles north of Sinclair, Wyoming, and still headed east. "The pilot reported he thought the ship had sufficient gas left to reach Des Moines [Iowa], and possibly further."[28]

At about four o'clock, the plane cleared the Rockies northwest of Cheyenne at thirteen thousand feet. Joey the kangaroo was unhappy in her crate and shaking her head from the ear-popping effects of altitude. At 4:15 a.m., the *Turtle* transmitted a message in Morse code to the Air Transport Command office at Denver: "V9082 B2B1 DAVIES 5 N XP 110 DM 19 DO375M CLRD to DM ESTD 60 N CX at 0414 M CDV."[29] To people on the ground, the text meant that Davies was flying at eleven thousand feet, at 190 miles an hour, about a hundred miles northwest of Cheyenne, and cleared for Des Moines. During the harrowing past few hours, the pilots had abandoned hope of reaching Bermuda, with Washington very much in doubt as well. The ATC passed along the message to the aircraft communicator at the Des Moines airport.

The *Turtle* found calmer conditions east of the mountains, but the weather remained much colder than back in Australia. The plane left Wyoming and flew into northwestern Nebraska and another time zone at 4:30 a.m. (5:30 central time). "Just prior to dawn, the flight commander said he was cutting speed to save gasoline, but that all was going well."[30] The sun rose over the Cornhusker State at 6:20 on Tuesday morning, October 1, the *Turtle*'s third dawn of the flight. During the coming day, reports of the crew's exploits would fill front pages and radio broadcasts across America—coverage the Neptune shared with bulletins from Nuremburg, Germany, where Hermann Goering and eleven other top Nazis were sentenced to the gallows for war crimes. Sometimes the *Turtle* was mentioned first, other times the condemned Nazis. "The Truculent Turtle flashed through the ether at a bad time for publicity purposes," an Illinois newspaper noted in an editorial.[31]

The plane radioed ahead to North Platte, Nebraska, that it would pass Grand Island at 6:45 a.m. and Omaha about forty-five minutes later. The *Turtle* flew south of Omaha at 7:30, as anticipated, and crossed the Missouri River below Council Bluffs, Iowa. From Omaha, the Green 3 airway turned slightly northeastward to Des Moines about 135 miles away and from there on to New York City. The great circle route to Washington, however, continued eastward.

A small crowd gathered at the Des Moines municipal airport in hopes of seeing a historic landing there. "The newsmen went on the air yesterday morning with the story that the Truculent Turtle, the Navy's long distance flight crew might land their plane in Des Moines. . . . KRNT's aviation reporter, having served three years as a B-29 pilot, was there to interview the fliers," a Des Moines newspaper reported.[32] About fifty cars of gawkers parked along the roadside as a knot of journalists and photographers gathered in the CAA communications office to wait for the Neptune. The local control tower had no direct contact but was monitoring the *Turtle*'s calls to Peoria in adjacent Illinois.

At 7:41 a.m., the fliers dashed hopes in Des Moines by radioing, "Point of intended landing is Columbus, Ohio. 430 gallons of fuel remaining."[33] Davies later said he had briefly considered a different landing point altogether. "I had a sudden impulse to land in Cleveland," he admitted. "I thought it might be nice to surprise the home folks."[34] But he suspected the navy wouldn't like it and continued flying toward Ohio's capital.

The *Turtle* left Green 3 at the bend and continued eastward toward the District of Columbia, passing twenty miles north of Ottumwa, Iowa, at 8:15 a.m. "Then, listening over the ATC phone, the local operations officers heard the tired pilot radio Columbus that if he found he had enough gas when he reached Ohio he would head for Washington, D. C."[35] Other Iowa airports were listening. Reporters and photographers rushed to the airport at Burlington along the Mississippi River in hopes of seeing the Neptune fly over shortly before nine o'clock, but the small city was miles off the *Turtle*'s path and too far away to spot it.

"Coming over Omaha and the Midwest today, we cruised at about 190 miles an hour," Davies said later.[36] The *Turtle* crossed the slender waist of Illinois in less than an hour. CAA employees in Moline listened to the *Turtle*'s communications as the plane passed over Peoria and continued toward sunny and frosty Indiana beyond. The aircraft communicator at the Purdue University airport had several messages to pass along to the *Turtle*, which approached West Lafayette around ten o'clock. "She succeeded in contacting the plane as it passed over the city and delivered the messages," a local newspaper reported. "She was informed that the plane was flying at 10,500 feet and was proceeding by way of Dayton, O., to land at Columbus, O." At such a height in good visibility, the reporter added, a plane would appear "merely as a speck in the sky."[37] The Neptune passed near Kokomo, north of Indianapolis, thirty minutes later.

Davies meanwhile radioed for the weather at Morgantown, West Virginia, on the western edge of the Allegheny Mountains, and at Washington DC. "Announcement of Davies' request for weather information was made by the Navy, which added unofficially that 'good tail winds' were reported over Ohio."[38] But the *Turtle*'s gas levels were alarmingly low, and Davies lost confidence in his gauges when one began to fluctuate, the needle bobbing from 250 gallons suddenly down to 75. "We didn't feel it was worth chancing a landing in a cow pasture between here and Washington when we already had a record," he said later.[39] As an Ohio newspaper observed in an editorial, "A landing at Washington would have been more spectacular. But Columbus was surer and safer."[40]

THE MUNICIPAL AIRPORT AT OHIO'S CAPITAL WAS CALLED Port Columbus, more in unrealized hopes of sounding glamorous than because it lay near any sizable body of water. The name seemed almost appropriate, though, during the war when the government established a naval air station there. The area also had another military airfield, Lockbourne Army Air Base, south of the city, but the *Turtle* aimed for the naval facility. Although the city was in the midst of a streetcar strike, a crowd had more time to gather there than back at Des Moines, and around fifteen hundred people assembled at the Port Columbus terminal. About 150 newspaper and radio reporters from as far away as New York City rushed in too. Twelve-year-old Tommy Torr pedaled his bicycle there to share the excitement. "I often visited the airport," he recalled decades later. "You could walk the flight line and look in the windows of all the airplanes. . . . No one ever questioned you; there were no fences."[41]

Columbus during the 1940s was a conservative and somewhat sleepy place, except when the Ohio State University football team played before tens of thousands of screaming fans in its huge horseshoe-shaped stadium. Residents' nickname for Columbus was "the Cowtown," partly because it was once the site of an enormous stock farm. Perhaps fittingly on this auspicious Tuesday, a small herd of cattle touched down at the airport ahead of the *Turtle*. "Seven prize Jersey cows were flown into Columbus yesterday in 12 hours from Ontario, Cal., arriving while a large crowd awaited the landing of the bomber Truculent Turtle."[42] The animals were there for a national cattle association show the following week.

A twin-engine Lockheed R5O Lodestar navy transport plane from NAS Glenview north of Chicago also arrived ahead of the *Turtle*. Aboard were ten newsmen and network representatives plus Rear

Adm. Edward C. Ewen, the current chief of U.S. Naval Air Reserve training and the future chief of navy public relations. The Lodestar touched down uneventfully in Columbus, but its brakes would fail during a second landing that night at Lambert Field in St. Louis. The plane went off the end of the runway, over an embankment, and onto railroad tracks below—all somehow without injuring the crew.

The Columbus skyline consisted of a lone skyscraper rising near the Scioto River. The forty-seven-story American Insurance Union Citadel, today known as the LeVeque Tower, rises six inches taller than the Washington Monument. The graceful tower had been a navigating landmark for biplanes during the late 1920s and early 1930s, and the *Turtle*'s pilots spotted it easily from dozens of miles away. Hundreds of the citadel's office workers eating lunch outside saw the Neptune sweep over downtown and the Greek Revival–style statehouse with its handsome round cupola around 12:20 p.m.

The airport was eight miles farther east, beside a vast Curtiss-Wright aircraft factory that had built over fifty-five hundred navy SB2C Helldiver dive-bombers during the war. A U.S. Marine Corps R5D ferrying officers, reporters, and photographers from Washington met the *Turtle* in the air as it neared the airport. The two planes twice circled Port Columbus together before the Neptune peeled off alone toward the southeast. As Davies then turned back for a landing, "one fuel gauge registered zero and the other 30 gallons," Reid said later.[43] "The big, blue, two-motored 'Truculent Turtle' circled in a steep bank at 2,000 feet and one engine sputtered," a wire service reported. "The crowd . . . watched anxiously in fear that the plane's tanks were running dry. Then, [Davies] leveled off the ship, the motor picked up, and he made a safe landing."[44] Davies later attributed the last-moment sputter to the Neptune's angle of bank rather than to fuel starvation.

With squealing tires and small puffs of smoke, the *Turtle* touched down at 12:28 p.m. eastern standard time. The crew officially had flown 11,236 miles nonstop—nearly halfway around the world—during fifty-five hours and eighteen minutes in the air, besting any mark the Japanese A-26 might have established during the war. Davies estimated that by deviating from their original great circle route, the *Turtle* actually had flown 300 miles farther. He said the Neptune had over 100 gallons of gas left in its tanks on reaching Columbus. Rankin later said there were 135 gallons in the main tanks and only 5 in the auxiliary.

The *Turtle* likely could have reached Washington if the forgotten 200 gallons had been pumped aboard at RAAF Pearce, although it would have had very little fuel left to spare. And had they received up-to-date weather data in the central Pacific and stuck to their original great circle track, the four fliers might well have continued out over the Atlantic. "Given favorable conditions, the trip to Bermuda would have been a cinch," project engineer Robert Bailey said in Perth.[45] He later added that the Lockheed technicians had expected the P2V to reach the island and were "disappointed when it didn't."[46]

The *Turtle*'s flight was nonetheless historic. "The purpose of our flight, more important than setting a new distance record, was to give the Neptune plane the acid test," Bailey said. "We wanted to—and did—prove how far a land-based patrol aircraft can go."[47] A *New York Times* editorial observed that the flight "emphasizes dramatically the huge distance we have come in aviation since the pioneering over-ocean flights of the Twenties, when the struggle across the Atlantic was a life and death gamble."[48]

The arriving Neptune taxied past the airport's art deco passenger terminal and up to one of the navy's functional humpback hangars nearby. The airport superintendent picked up a telephone to call the newsroom in one of Columbus's three daily newspapers. "We've got something kind of unusual out here," he said with midwestern understatement.[49] The fliers were surprised by the size of the crowd. Uniformed navy shore patrolmen struggled to hold back a surge of spectators as photographers snapped away, and a mobile unit from a local radio station broadcast the *Turtle*'s arrival. Sadly, a mother near Wilkes-Barre, Pennsylvania, would see one of the published photos and believe she had seen her missing son among the sailors. Her boy was lost during the war when his ship sank off the coast of New York. The woman contacted the Columbus police and asked them to check, but her son wasn't listed among the naval air station's personnel.

The *Turtle*'s fuselage showed streaks of engine exhaust after its long flight, but the crewmen had spruced themselves up before landing. Climbing down onto solid land, the four wore crisp, forestry-green aviation winter working uniforms and were "clean-shaved and fresh as the proverbial daisies."[50] Waiting to greet them were Admiral Ewen, Capt. John Foster of NAS Columbus, Mayor James A. Rhodes, and a horde of journalists. "I'll give you two minutes!" a navy official shouted at the press before proceedings moved inside one of the large hangars.[51] Someone lifted the kangaroo's crate out of the plane and opened the door. Little Joey was "pretty well tired out," Davies said, and she promptly fell asleep on the concrete apron.[52] Mayor Rhodes, who later went on to serve four terms as the state's governor, "made a pitch for the kangaroo for the Columbus Zoo but was turned down."[53]

A senior navy medical officer flown in from NAS Anacostia tested the crews' eyesight, blood pressure, respiration, and reactions, and pronounced them all in "fine shape."[54] Tabeling later wrote that the crew had thought the *Turtle* would be "the biggest guinea pig," but he found they all were test subjects as well.[55] The four aviators also were temporarily done flying, grounded over concerns about fatigue. Two-plus days in the air were enough. Their assignment now was navy public relations.

"After coffee and cigarettes, the first smokes since the flight began, the crewmen hustled through a wild press conference which lasted more than an hour," the AP reported.[56] A pool table in an officers' lounge served as a makeshift dais, and the fliers looked fresher than the gaggle of reporters who had been up all night chasing the story. Davies sat on a chair hoisted onto the green baize where striped and colored balls normally clicked and clacked. Reid, Tabeling, and Rankin stood behind him on the table as Davies answered questions from the reporters clustered in front of them. The commander first hailed the *Turtle,* which he called "a pilot's dream."[57] He assured everyone that the crew had come through in good condition. "It was a swell trip," Davies said. "I'm not very tired. I got 12 to 15 minutes sleep during the trip. We all took turns caring for the kangaroo."[58] The pilot patiently guided the reporters through the facts of the long flight and its many challenges.

Rankin's parents in Oklahoma heard him talking with broadcasters on the radio. A network also arranged a long-distance phone call for the fliers with their wives in Washington. The three married men broke away to cluster around a single black telephone. "I'm so proud of you I don't know what to do," Virginia Rankin told her husband. "We're happy that you're down," added Eloise Davies, who had gone to bed at two in the morning only after learning Tom's plane had safely reached America.[59] Walt Reid asked his wife for a steak dinner when he reached Washington. "Bring along that kangaroo," Ruth Reid cracked. "It's the only kind of steak you'll get in Washington."[60]

The fliers spent a little over three hours in Columbus. Despite earlier denials of weariness, Davies said, "I'm going to go to sleep and not get up for two days."[61] The *Turtle*'s crew left at 3:55 p.m. as passengers on the R5D that had come over from Washington, with Tabeling's grocery-bag luggage somehow going astray during the transfer. A fresh pilot and copilot from the Naval Air Test Center at NAS Patuxent River, Maryland, flew the refueled Neptune the last few hundred miles to the capital.

A *Life* magazine photographer captured the *Turtle*'s fliers relaxing in the transport, although his black-and-white photos never made it into the magazine. The four men chatted, smoked cigars, and read a Columbus newspaper, with Rankin donning an Australian slouch hat, a souvenir from Down Under. The crew landed at 5:20 at NAS Anacostia, where navy brass were assembled to greet them. "As they bunched on the cement apron in the nippy dusk, the admirals and captains told each other that 'we beat the army,'" a Cleveland newspaper reported.[62] Rankin later told his parents "that as a surprise to them [the] first thing they saw upon alighting at the airport was the 'Truculent Turtle' which had been flown on to Washington by another crew to its share of the ovation."[63]

Also awaiting the fliers were several enthusiastic well-wishers, "the happy embraces of their wives and Distinguished Flying Crosses from Secretary Forrestal of the navy."[64] The married men tumbled out onto the apron first, followed by bachelor Tabeling, who "grinned happily at the family knot," the United Press reported. Davies embraced his wife, did it again for the print cameramen, and then a third time for the newsreel boys. "Finally a navy officer broke up the mob scene and cleared the way for Forrestal."[65] The navy secretary in a gray civilian suit and fedora chatted briefly with the fliers and their families. "It was no tougher than a good long patrol mission," Davies said of the hop when presented with his award. "Well, Commander," Forrestal replied, "in our book, flying 11,236 miles non-stop from Perth to Columbus really rates a medal."[66]

A newspaper in Columbus applauded the navy's recognition. "Credit is due the crew which made the record possible, for theirs was a long and tiresome vigil during which the airplane crossed the Pacific the long way, flew over the islands where war raged only 13 months ago and went on to cover more than half the United States."[67]

Following the ceremony, reporters and photographers turned their attention to Joey, who wasn't fated to become anyone's supper after all. An assistant head keeper from the National Zoological Park collected her in the crate. Broadcaster Lowell Thomas described her that night as the "truculent kangaroo aboard the Truculent Turtle."[68] But Walt Reid later wrote that Joey "regained her spirits and posed for pictures very nonchalantly with the director of the Washington Zoo."[69] Visitors would see her in the zoo's small mammal house the next day. Photos of the globe-trotting marsupial appeared in newspapers across the country, with the *Christian Science Monitor* declaring that "among the race of kangaroos, Joey must surely rank as the greatest bounder of them all."[70] The *Oakland Tribune* ran a poem on its editorial page:

On the Truculent Turtle a kangaroo
Journeyed thousands of miles without a stop
Thus settling a question, emphatically too,
"How far can a kangaroo hop?"[71]

The *Turtle* and its photogenic kangaroo had beaten the *Dreamboat*, the AAF's bigger, flashier, and faster hare, across the finish line in the race to capture American public opinion. But that didn't mean Colonel Bill Irvine and his B-29 crew had given up on reaching Cairo. To the contrary, the army's long-distance dash over the Arctic promised to be as spectacular as the navy's lope across the Pacific.

FOURTEEN

The Pole

BILL IRVINE SENT A PERSONAL MESSAGE TO TOM DAVIES that he was "very proud" of the navy fliers and their Neptune. "You and your crew have done a magnificent job of demonstrating the outstanding capability of American pilots and aircraft," the *Dreamboat's* commander wired from Honolulu. "Your planning and execution must have been perfection itself. Sincerest congratulations to you and the Truculent Turtle crew from me and the Dreamboat crew."[1] Davies returned the courtesy with a nod toward the AAF's planned hop over the Arctic. "Best of luck to you and your crew," he wired back to Hawaii. "I know the results of your flight will be a signal contribution in the field of aviation."[2]

Irvine conceded that his rival's enviable success might change plans for the *Dreamboat*. He knew his Superfortress couldn't achieve its maximum range during the flight to Cairo because it had to fly too high to get there. "My first reaction when I heard the Turtle had passed Des Moines was that the navy boys had done a terrific job and that they had got a lot out of their plane," Irvine said. "Now it turns out that it was a better plane than one might have suspected."[3]

On Tuesday the weather officers at the AAF Master Analysis Central in Washington notified Captain Yorra in Honolulu that Friday,

October 4, looked best for beginning the flight across the Arctic. Irvine confirmed to reporters on Wednesday that the *Dreamboat* definitely would go ahead but didn't give a timetable. "Although Dreamboat crew members were admittedly disappointed over the loss of incentive to set a new nonstop record, they contended the real purpose of the flight is to acquire new knowledge of polar flying conditions and to test new navigation instruments," the *Star-Bulletin* reported.[4]

"The strain of getting ready was probably the worst of all," navigator Major Brothers recalled. "We spent nearly five weeks in Honolulu waiting for the right conditions to make the trip."[5] The crew awaited "favorable weather reports before starting the flight," the *Honolulu Advertiser* said Thursday morning, when the record flight of the *Truculent Turtle* was still on everyone's mind.[6] "The point is that with fair luck the navy plane has covered a distance which is probably beyond the reach of the Dreamboat under any but the most favorable circumstances," a *Star-Bulletin* editorial noted later that day. "And, of course, this is exactly what the navy crew was driving for." Despite the long odds, the paper was "pulling for the Dreamboat to turn the tables."[7]

Everything went well on the *Dreamboat*'s test flight Thursday morning. Irvine and Yorra held a conference call at 4:30 that afternoon between the Hickam weather office and the Master Analysis Central at the Pentagon. The pair said the *Dreamboat* would begin the flight to Cairo the next day "unless there was an unexpected deterioration in actual weather conditions, or in the outlook, prior to that time."[8] Immediately afterward, Irvine sent an alert to every communication, weather, and navigation check station along the route between Oahu and Cairo.

All that night, Boeing engineer Lowell Houtchens and Major Brothers conducted preflight inspections on the *Dreamboat*. The ground crew meanwhile pumped thirteen thousand gallons of fuel into the bomber, inserting one-third before they towed it from Hickam Field over to the adjacent airport. It took six hours to get all the gas aboard, and afterward the *Dreamboat* weighed over seventy-four tons.

The *Dreamboat*'s crew had much different histories than the *Turtle*'s. None of the AAF men was a West Pointer, and four were what the navy called mustangs, or former enlisted men who had become officers. They were an ambitious, self-made, and gung-ho bunch intent on reaching Cairo. No one slept much in the barracks that night. Hickam Field was showing the movie *The Strange Love of Martha Ivers* with Barbara Stanwyck and Van Heflin, and the soundtrack floated in through open windows to keep the Dreamboaters awake. The fliers rose early and left their quarters at three o'clock that Friday morning. Dick Snodgrass was making the hop with them. The Boeing employee seemed sent by central casting to play a bespectacled flight engineer, and the crew recognized his value. Master Sergeant Vasse called Snodgrass "the mathematician who figured it out and made it possible."[9]

Colonel Irvine and his crew breakfasted on eggs, toast, and coffee at the ATC terminal. "Then in the drowsy darkness preceding dawn a line of 10 autos carrying army officials and reporters snaked out of Hickam field, weaved toward John Rodgers airport and cut across the landing strips to where a silvery B-29 stood in a semicircle of floodlights," the *Honolulu Star-Bulletin* reported.[10] The gleaming Superfort had the PACUSA insignia painted on its tail and "an identification orange and black No. 7 under its right wing. . . . On either side of its nose is inscribed 'Pacusan Dreamboat, Boeing Superfortress,' with the names of the crew painted alongside in a rectangular box."[11]

A navy nurse draped leis, floral gifts from the Hawaii Visitors Bureau, around the fliers' necks and over their yellow Mae West preservers as photographers' flashbulbs popped and hissed. The *Dreamboat*'s officers synchronized their watches with a navy captain, the flight's official timer. Irvine chatted with Don Whitehead of the Associated Press and other reporters with a box of cigars tucked under his arm. "I've sweated out combat missions which were lots worse than this," the colonel said with a grin. "All the boys are feeling fine about the trip. We are confident we can make it in about 41 hours."[12]

Like the *Truculent Turtle*, the *Pacusan Dreamboat* carried local souvenirs on the long hop: twenty-five copies of an Egyptian special edition of the *Honolulu Advertiser* plus a letter to the country's monarch from the newspaper's president and general manager. A 120-point headline filled the front page above the fold: ALOHA TO KING FAROUK AND PEOPLE OF EGYPT FROM HAWAII VIA THE PACUSAN DREAMBOAT. "'It'll be a pleasure to deliver the Advertisers,' Col. C. S. Irvine, the AAF 'paper boy' for the edition, told Raymond Coll, Advertiser editor, when he received the bundle."[13] The colonel probably was delighted that nobody had to look after or feed the bundle, which was lighter and occupied less space than a crated kangaroo. Major Hays was a courier, too, "in charge of a package containing a Hawaiian aloha shirt to United Press Correspondent Walter Collins, in Cairo, from United Press friends in Honolulu."[14]

Former marine Keyes Beech of the *Star-Bulletin* talked with the crew before they climbed into the plane. Irvine mentioned the *Dreamboat*'s earlier bad luck, admitted that the *Turtle*'s flight had been a "kick in the pants," and worried about encountering bad weather and maybe ice over Alaska. Navigator Hays joked about the high Alaskan and Canadian mountains, and how Irvine would be mad if they didn't fly between them. Master Sergeant Fish, the young combat-hardened crew chief, inspected the engines a final time and found everything in top condition. Fish conceded, however, that despite hopes to the contrary, the B-29 would fall a thousand miles short of beating the *Truculent Turtle*'s nonstop distance record. But everyone was still eager to get away, with wise guy Major Kerr quipping that Shepheard's Hotel in Cairo would be annoyed if the AAF canceled its reservations again. "Goodbye all," Kerr cracked. "We'll probably be back in 20 minutes."[15]

The last fuel truck pulled away as a tractor hooked onto the *Dreamboat*'s nose wheel and tugged the B-29 around into takeoff position. An irreverent reporter called out a final question—did they favor statehood for Hawaii?—making the crew laugh. Then the floodlights clicked off, the electrical generator went quiet, and the journalists and spectators drifted away. Soon they regathered at the eight-thousand-foot mark of the combined runways, where the bomber was expected to lift off.

Copilot Bev Warren was better than anyone in the AAF at yanking a grossly overloaded Superfortress into the air, so he was handling the *Dreamboat* during takeoff. Earlier in the week, Warren had dropped one rank to lieutenant colonel as temporary wartime ranks fell back to Earth during peacetime; in the cockpit he still wore a bird colonel's spread-eagle insignia anyway. The B-29 had a third pilot in addition to Irvine and Warren. Former test pilot Maj. James R. Dale of Wise, West Virginia, had flown five combat missions over Japan and was aboard the *Dreamboat* primarily as a flight engineer.

Irvine settled into the pilot's seat opposite Warren, happy to let the laconic Texan take the controls during the first crucial moments of the transpolar flight. The colonel was a little boy in Nebraska when the Wright brothers first took to the air from North Carolina sand dunes in 1903. "I know what they're saying about me," Irvine had joked after an earlier postponement. "They're saying, 'Why doesn't the old so-and-so get the hell out of here?'"[16] Now the cigar-chomping colonel was set to depart at last on what he had promised his family back in St. Paul would be his last long-distance record attempt.

The *Dreamboat* was taking off downwind to avoid downtown Honolulu and the mountains beyond, but tailwinds weren't a concern at only four miles per hour. Test flights had shown that propeller torque affected the overloaded B-29's handling on the ground, so the pilots would use their brakes while accelerating down the long runway. The braking would cost them speed, but Irvine and Warren had concluded they could get the plane safely airborne at about seventy-five hundred feet. The colonel had ordered markers posted every two thousand feet to indicate their progress. They could abort takeoff until six thousand feet if the plane lost one of its four engines.

Beyond that point, they would have to struggle up into the air, dump fuel from tanks in the bomb bay, and fly around for an hour, burning more fuel, before returning to the ground twenty-five thousand pounds lighter.

The sky began to lighten at 5:20 a.m. Twenty-five minutes later, Warren had the propellers whirling and the bomber pointed down Runway 26. After a month of delay and frustration, they were ready at last. "Here we go!" Warren shouted at 5:49.

He pushed the throttles open, and the *Dreamboat* began moving, its engines roaring at twenty-eight hundred revolutions per minute. Warren pulled back lightly on the controls at the first two-thousand-foot marker, and the nose wheel left the runway. "I just held it there until the plane flew itself off," he said later. The bomber lifted free after sixty-five hundred feet at 5:51 a.m., doing nearly 150 miles per hour. Warren turned to the colonel. "Bill," he said in relief, "this is no different from a thousand other take-offs we've made together." Irvine replied, "If you ever had a take-off in your system, this was the time to do it."[17]

Keyes Beech stood with the reporters watching the Superfort grow small in the distance. He wrote for the *Star-Bulletin*, "The great Superfortress headed into a blue gray dawn and out over the blue sea just as the clouds began to break over the Koolau mountains and the sun crept up over Diamond Head."[18]

DREAMBOAT FAMILIES FOLLOWED THE FLIGHT AS CLOSELY as they could on the mainland. Captain Saltzman in Washington notified Colonel Irvine's brother George at the family farmhouse in St. Paul that the B-29 had taken off. "Mr. and Mrs. Irvine, their son Billy, and the pilot's sister, Mrs. Charles Petersen of Sherman Oaks, Cal., took turns sleeping so the radio would be manned at all times."[19] Lieutenant Colonel Warren's wife, Beatrice, also waited out the hop in Nebraska at the home of her father, Stanley Beranek, a prominent Omaha druggist. "During the flight Mrs. Warren and their two-year-old son, John Beverly, who is the 'apple' of Grandfather Beranek's eye, were able to keep in fairly close touch with developments by means of radio teletype messages made available to them by an Omaha radio station."[20] And Major Kerr's wife, Colleen, waited in Nebraska too, in her tiny hometown of Bayard on the western side of the Cornhusker State.

Other *Dreamboat* wives, parents, and siblings followed the newspapers and radio bulletins in homes scattered from Southern California to the Northeast. Newlywed Mildred Fish waited for news at her mother's home in Pittsfield, Massachusetts. She'd married the *Dreamboat*'s crew chief in June, five days before he reported for duty on the *Dreamboat*. Mrs. Mary Bullett approved of her son-in-law. "He's a nice boy," she said later. "He's got a lot to brag about but he never has much to say."[21]

Her daughter was a former WAC enlisted woman who had met the sergeant at an army airfield in Georgia. Mildred Fish was a "particularly enthusiastic rooter" for the Cairo flight, according to a local newspaper. "She last saw her husband in July in Oklahoma City, before the Dreamboat left for Hawaii. She plans to meet him in New York at the completion of the return flight."[22] A paper in nearby Albany, New York, added that Mildred was "watching the progress of the flight closely but 'wasn't worried about the outcome.'"[23] She admitted later to sleeping on a couch near the radio, with an alarm clock set to ring every hour so she could tune in for bulletins.

THE *TURTLE* HAD DEPARTED AT DUSK, BUT THE *DREAMBOAT* took off at dawn. The B-29 headed over the Pacific, flying not much faster than the Neptune had to conserve fuel. Irvine and Warren flew the first five hours at a thousand feet in excellent weather. Major Hays initially used dead reckoning because the plane was so low over the Pacific that he could judge the wind effect and therefore the drift by watching the wave action below. When eventually they hit a weather front, the fliers slowly climbed up through the soup to over five thousand feet. The navigators used their radio compass to check their position when passing over the weather ship designated Dog, which had supported the *Truculent Turtle*'s flight. The plane maintained its altitude almost until reaching the southeastern coast of Alaska, twenty-seven hundred miles northwest of Honolulu.

Through an Anchorage radio station, part of the Alaska Communications System, people in Sitka followed the *Dreamboat*'s progress. The local newspaper reported that the bomber was expected overhead about seven o'clock that evening Alaskan time, which was ninety minutes ahead of Hawaii's. That afternoon a Boeing radio monitor picked up a message from Irvine, who "requested that the Army send B-17 [*sic*] several hundred miles inland from Sitka to provide latest reports on weather conditions. He asked that the planes fly over his route leaving Sitka about two hours before his expected arrival here."[24]

Lieutenant Colonel Shannon and M.Sgt. Edward Vasse transmitted by code and voice to their far-flung network of amateur radio operators. Vasse first napped for a bit following takeoff, having stayed up all night to help fuel the plane. He and Shannon were an odd pair, a quiet young noncom and an extroverted silver-haired officer, respectively. Vasse had installed their standard army-navy AN/ART-13 radio himself and had spent endless hours fiddling with it since Oklahoma City, testing and retesting, checking and rechecking. The set was designed for use only twenty minutes per hour, but Vasse figured they could operate it pretty much continuously during the flight. He and Shannon did almost exactly that as the *Dreamboat* winged toward North America.

George W. Spare (call sign KH6CT) in Hawaii was their primary ham radio contact early in the flight. "After the take-off and just prior to it, it was something to behold . . . hams removing other hams from my frequency (14,204 kc.)," Spare said later. "The frequency was kept clear for about two-and-one-half hours . . . *something*, we'll all agree, in the congested 100-kc 'phone band."[25] The *Dreamboat* transmitted on ten different frequencies, depending on time, location, and atmospheric conditions. The hams used arcane acronyms, codes, and phonetic aids to communicate. An operator whose call sign was

W1EH, for instance, would identify himself (nearly all were men) as William-One-Edward-Henry. They also used old railroad telegraphers' codes, such as *30* to signify the end of a transmission, the same notation newspaper reporters added at the end of their copy.

The *Dreamboat* was over the North Pacific Ocean when Shannon (W3QR) spoke for thirty-five minutes with Joseph Schobert (W7CGL) and Louis Huber (W7CRJ) of Zenith, Washington, during the longest continuous amateur contact of the flight. The two hams were Boeing employees who got to know Shannon while the bomber underwent preflight preparations in Seattle. Schobert relayed the gist of the message to reporters: "Everything progressing well, having fine trip."[26]

Sergeant Vasse's friends and family back in Huntsville, Missouri, called him by his middle name, Gray. Irvine praised the former shoe clerk later as "one of the Army's best radio instructors. And one of the finest radio mechanics in the business."[27] The long flight over water must have taken on a strange dreamlike quality for the slender airman, who had been lucky to survive the war. Vasse enlisted in 1939, the first of three brothers who served in uniform. He helped ferry bombers to embattled England in 1940 before America entered the war. Later, the Missourian was aboard a troopship bound for the Philippines when Japan attacked Pearl Harbor.

"His squadron had been rushed to Java to bolster Dutch forces and guard against invasion there," his hometown newspaper reported. "They arrived at Java just before the fall of Singapore, but bombed their way out of encroaching invasion and were among the first to meet the J— in aerial conflict."[28] Vasse flew on bombing raids and saw aerial combat with the Nineteenth Bomb Squadron. He especially remembered the day Mitsubishi Zero fighters jumped his Consolidated LB-30 Liberator somewhere over the Celebes.

"I think it must have been ten years, but it couldn't have been more than 20 minutes," he said. "Scared? My knees turned to water. I was absolutely scared green and I'm not a bit ashamed of it. . . . After those guns get to knocking, you're O.K."[29] Vasse had sewn a master sergeant's three chevrons and three rockers onto his sleeve in 1943 at age twenty-four. Now in 1946 he was riding in a famous B-29 bound for the top of the world and beyond, working at the radio operator's station beside a garrulous light colonel who sat chatting with civilians, without a Japanese fighter plane in sight. Such was life in the AAF.

THE *DREAMBOAT* WAS STRIPPED OF ALL NONESSENTIAL equipment to reduce weight, but it flew across the North Atlantic well equipped with survival gear should it go down over land or sea and especially if it crash-landed in the Arctic. The gear included a one-man life raft attached to each parachute plus two seven-man rafts and signal flares. In addition, the plane carried canned and concentrated food and juices to last three weeks, and the crew had received instructions on building igloos and shooting polar bears, seals, and walruses. Also stowed aboard were Coleman lanterns, a gasoline stove, fishing tackle, cords for snares, ice axes and ordinary axes, ten carbines with ammunition, and a device for desalinating seawater. Each man had mukluks, four pairs of socks, a sheepskin jacket and pants, three pairs of gloves, a parka, snow glasses, and two sets of snowshoes. All this equipment weighed about 120 pounds per person. But the crew wasn't worried about survival. They thought the world of their young crew chief and trusted him to keep the *Dreamboat* flying. "We're more proud of Gordon Fish than any member of the crew," Irvine said later. "The lad really knows his B-29."[30]

The son of an accountant father and immigrant Norwegian mother, Fish had joined the army in 1940 and served in the Pacific during the war, rising like Vasse to master sergeant. For a while the Appleton, Wisconsin, native had served as the crew chief for Gen. Jimmy Doolittle. Fish reenlisted the same week he married in June "just so he could go on the trip," his bride declared. "He loves flying. He eats and sleeps it."[31] Rather than take his ninety-day reenlistment furlough, Fish worked furiously to get the Superfortress ready for this epic flight, losing twenty-five pounds during the effort. He constantly watched for gremlins and hidden problems on his planes. "Once you learn to expect trouble it's not so bad," Fish said years later. "It's not like driving down the street expecting no trouble and then having a blowout—things that come unexpected like that are bad."[32]

Nearly twelve hours into the flight, at 5 p.m. Alaskan time, Lieutenant Colonel Shannon transmitted from above the North Pacific Ocean. Hams Schobert and Huber heard the call and sensed anxiety in his voice. "This is Army 4061, the B-29 Dreamboat," Shannon said. "We're in a frontal area and we want to know in a hurry how to climb out of it. We're trying to contact one of our B-17 weather ships."[33] A plane designated WX-2 replied with a detailed weather report for the area: 70 percent stratocumulus cloud cover at Sitka, a thin layer of altostratus clouds over the coastal areas, a stratocumulus undercast in the vicinity of the Stikine Mountains in British Columbia, cloud decks reaching no higher than eleven thousand feet and extending at least seven hundred miles into Canada, and twenty miles' visibility with no ice or turbulence. WX-2 also supplied temperature, wind direction, and more. No doubt relieved, Shannon arranged for the B-17 to scout ahead as far as Coppermine on Coronation Gulf near the Beaufort Sea.

Darkness fell before the *Dreamboat* reached the Alaskan coastline. Clouds parted enough for the crew to see a few lights at Sitka below. The navigators set a course for Coppermine using their directional gyro, which they checked with the astrocompass as soon as they could see stars. About ninety miles beyond the town, they caught a glimpse of Alaska's capital, Juneau. The weather plane flying ahead had assured the pilots that the *Dreamboat* wouldn't pick up any ice. But conditions can change quickly in the air, especially over Alaska, and contrary to the prediction, the *Dreamboat* began to accumulate rime ice. This sort of ice is formed by supercooled water droplets freezing on contact with the leading edges of airfoils and engines. Fortunately rime ice is brittle and doesn't disrupt lift as much as clear ice. It caused no significant problems over Alaska for the *Dreamboat*.

Ahead lay the "most imposing range on the continent," the towering St. Elias Mountains, which rose like medieval fortifications beside the gravel-surfaced Alaska (Alcan) Highway that the U.S. Army built during the war.[34] Beyond the St. Elias and other ranges lay the Mackenzie River Basin, three times the size of France. "The crew broke out altitude clothing and oxygen masks for the mountain crossing," the United Press reported, "as Canadian Arctic radio operators remained alert [all] night long in their isolated snow-drifted huts along the fringe of the arctic ice pack and on the bleak, low islands between the mainland and the north pole."[35]

Major Kerr might well have wondered how he had come to be flying over so cold and forbidding a landscape after spending so many months in the Pacific. Born in New Mexico and reared in sunny Southern California, Kerr had hoped to become a banker before being drafted in 1941. He rose to sergeant in the infantry before shifting to the AAF, earning a commission, and rising quickly again thanks to his keen administrative abilities. The cheerful wisecracking major had served with Bill Irvine longer than anyone on the *Dreamboat* crew and was variously listed as an aircraft engineer, scanner (watching for engine fires), personnel and equipment officer, and observer.

"But beneath Kerr's flip exterior there is a shrewd mind quick to size up a problem and cope with it," the *Honolulu Star-Bulletin* said.[36] On the ground, he was Irvine's executive officer and wizard troubleshooter; he'd flown from Honolulu to Los Angeles, for example, to supervise the repairs and inspection of the troublesome fuel tank. But in the air now, with a very long way to go before reaching Cairo, Kerr had to trust his crewmates and Dick Snodgrass from Boeing to perform their duties as consummate professionals.

The *Dreamboat* quickly crossed the narrow strip of southeastern Alaska to enter Canadian airspace in northwestern British Columbia. The bomber flew within fifty miles of Whitehorse in the southern Yukon, making a radio check before starting across the polar region. Whitehorse was where the RAF *Aries* crew had landed following their polar flights during the war; it was also the terminus for the recently abandoned Canol Pipeline, through which vital oil had flowed from oil fields in the adjacent Northwest Territories. Ahead lay more mountains, Great Bear Lake with the isolated uranium mining village of Port Radium on its eastern shore, and thousands of smaller nameless bodies of water scattered across Canada. Past all that stretched the Arctic Circle, beyond which there were days during the year of no darkness and other days of no sunlight. Farther still were Coppermine and other remote places before finally the plane would near the magnetic North Pole somewhere in northern Canada. The fliers perhaps took some comfort from knowing that a string of tiny, twelve-watt Royal Canadian Mounted Police radio stations on Victoria, Somerset, and Baffin Islands were listening and hoping to make contact.

The crew pulled on their electrically heated flying suits to avoid using the heating system and save fuel as temperatures fell inside the *Dreamboat*. "We found that the Polar region is colder on the ground, where the temperature averaged 50 degrees below zero [Fahrenheit] than in the air, where, flying at 17,000 feet the temperature is 30 degrees below zero," Irvine said later.[37] Master Sergeant Fish echoed the colonel's comments after the Burbank-to-Brooklyn run in December 1945, remarking that the flight to the Far North was "smooth as silk, but cold as hell."[38]

Readers and listeners around the world sat transfixed by frequent newspaper and radio bulletins describing the bomber's progress. The transmissions from Shannon and Vasse, "the terse radio messages, afford thrilling reading in a day when headlines and loudspeakers intercept adventure in the act," the *Christian Science Monitor* observed. "McKlintock [*sic*] Channel, Victoria Land, Boothia Peninsula, Baffin Bay, Greenland Icecap—the names of these remote polar regions make almost hourly landfalls in the 250-knot stride of the Pacusan Dreamboat."[39]

The nearer the *Dreamboat* flew to the magnetic North Pole, the more Major Hays's instruments began to misbehave. "My compass went round and round," he said. "I got so tired looking at it that I covered it up."[40] But his celestial navigation equipment worked as expected, and with clear skies in the high Arctic, the navigators could comfortably rely on dead reckoning as well. At 9:00 a.m. Greenwich mean time (4:00 a.m. eastern standard time), after nearly fifteen hours in the air, the *Dreamboat* was flying northeasterly on a true heading of seventy degrees. But time hardly meant anything this far north, where all zones converged on the geographic North Pole.

Weather conditions were excellent for both flight and navigation, and the crew were able to look straight up at Polaris, the North Star. Hays took a star shooting on Altair, Rigel, and Dubhe while over the heart of the troublesome magnetic area and determined his position as 70 degrees 45 minutes north, 103 degrees west, off the eastern coast of Victoria Island. As the *Dreamboat* neared the magnetic pole, the fluxgate and pilot's compasses grew increasingly sluggish, swinging through 30 to 40 degrees within a couple of minutes. Listeners picked up a radio message at 4:45 a.m. that the crew was watching a "spectacular aurora borealis" and reporting, "Everything okay."[41] The plane then entered the blackout zone, caused by the aurora, sunspots, and ionospheric conditions. According to ham operator Huber back near Seattle, this combination "effectively silenced the *Dreamboat*; and for the next five hours kept it silent."[42]

Three minutes after the bomber's terse transmission, Hays made another celestial check using Rigel as his reference point. When the gyro's heading didn't agree with the figure he expected, he rechecked using Rigel, Pollux, and Dubhe. Based on the position of these three stars, he calculated the plane's position as 72 degrees north, 85 degrees west, over western Baffin Island. At some point during the forty-eight minutes between making his first and second triple-star checks, the B-29 passed within "a few miles" of the magnetic North Pole.[43] Hays later estimated that the wandering landmark lay north of the plane's course, rather than to the south as anticipated, and 250 miles to the left of the spot marked on his charts.[44] "If either Great Britain or the United States would take the trouble to explore, it would probably be found that the South Magnetic Pole is equally elusive," the *New York Times* science editor correctly predicted later.[45]

Hays was very satisfied with the performance of his instruments despite the anticipated variations in readings along the upper limits of the flight across the frozen north. "Actually, the trip over the pole at this time of year is a celestial navigator's dream," he said. "The flight probably is the most difficult in the world from both navigation and weather points of view, but we flew over the clouds and had all the celestial bodies in plain view."[46]

Master Sergeant Vasse managed to overcome the radio blackout amid extraordinary atmospheric conditions. He tapped his key as the B-29 flew over the northern tip of Baffin Island, communicating in dots and dashes with an operator in Hawaii "as plainly as if the phone were in the same room."[47] A little later, the RAF Fighter Control base at Prestwick, Scotland, picked up a transmission too. The *Dreamboat* crew meanwhile gazed down on a landscape resembling Norwegian fjords and icebergs that reminded them of wartime ship convoys. As the bomber began the long "downhill" segment of the flight, it would have been comforting for the men to think that the hardest part of the flight was behind them. But there were no easy hours during the long hop over the Arctic between Honolulu and Cairo.

FIFTEEN

Downhill

MAJOR HAYS KNEW THE *DREAMBOAT* HAD ONLY NINETY minutes of daylight above the Arctic Circle. He had advised the pilots to watch for the sun to rise slightly to the right of the B-29's heading, so it was a jolt when Bev Warren called back to him, "Hays, you better come up here and take a look at the sun. It's coming up about 20 degrees to the left and that's not what you told us to expect." The navigator rushed to the cockpit, wondering what had gone wrong. Had they somehow strayed badly off course?

Hays soon realized with relief that he'd been tricked by the unique characteristics of flying in the Arctic. The sun there traverses the sky very quickly compared with locations farther south but takes much longer to rise. As the bomber flew over eastern Baffin Island at official sunrise ninety minutes later, "Old Sol" lay five degrees to the right of the nose—exactly where Hays expected it. "Had you scared, didn't we?" joshed Warren, who'd never really worried about their position.[1]

Sunrise also brought a rare and spectacular natural phenomenon caused by ice particles in the air acting like a prism on early morning sunlight. "The 'green ray,' as this curious coloring is called, can be seen at sunset or sunrise when the sun's rim is just above the horizon," the Air Weather Service explained.[2] With the landscape visible below, Hays precisely charted his position as the *Dreamboat* pushed on and put Canada on its tail. "Ahead stretched the bleak ice floes of Baffin Bay before the plane reaches the west coast of Greenland, the last big hop across land before the ship reaches Europe," the *Brooklyn Eagle* told readers.[3]

At 9:45 a.m. eastern standard time, almost twenty-two and a half hours into the flight, the ATC base at Narsarssuak (Narsarsuaq) at the southern tip of Greenland picked up a transmission saying the plane was over the Greenland icecap. About an hour past the coast, the bomber began bouncing through storms and high headwinds, an unwelcome surprise that Hays later attributed to misinterpretation of data in a forecast passed along to the crew. The poor conditions forced the navigators to work for a while by dead reckoning. The *Dreamboat* left Greenland and flew out over the Denmark Strait bound for Iceland.

"The Americans are having a warm love for the cold countries: The Arctic, Iceland, Greenland and Scandinavia," the Soviet newspaper *Izvestia* commented.[4] Navigator Brothers gave his account to a Tennessee newspaper: "From Hawaii to Iceland the trip 'was a breeze' the major said. The bomber ran into ice as it neared the Icelandic coast. Since the B-29 carried no de-icing equipment, Brothers explained, they 'went upstairs' about 21000 feet."[5] By noon eastern standard time (4 p.m. local time), the *Dreamboat* was flying over the volcanic island that during the war was a vital refueling stop for planes ferrying men and equipment to the United Kingdom. While above Iceland, Frank Shannon picked up an AAF weather reconnaissance plane whose call sign was WX-2, the same as the B-17 back at Sitka. The identification puzzled Shannon because when he'd last heard of the plane, it was landing back at Whitehorse.

Observing communications security, this second WX-2 said only that it was part of the Fifty-Third Weather Reconnaissance Squadron headquartered at Newfoundland.[6] This secretive squadron began experimental weather flights in Wisconsin during the wartime summer of 1943. A reporter who caught a ride on one of its Flying Fortresses in the spring of 1947 would call it "the U. S. Army's pioneer weather-checking air outfit." He wrote, "Flight B wrote Arctic history last fall by running interference for the Pacusan Dreamboat along a route from northeast Baffinland, over the Greenland ice cap and Iceland nearly to Scotland."[7] Later, while operating out of Mississippi, the Fifty-Third became famous as the Hurricane Hunters, a role it fulfills today as a reserve squadron. Shannon patched WX-2 through to the cockpit so Irvine could ask its weathermen directly about conditions ahead.

The *Dreamboat* took about an hour to clear Iceland. London lay nearly twelve hundred miles southeast of Reykjavík. At 8 p.m. Greenwich mean time, the bomber approached the Outer Hebrides and northwest Scotland. The British Air Ministry planned to have de Havilland Mosquito fighter-bombers from RAF Northolt greet the *Dreamboat* and escort it down across the capital. The Mosquitos would drop red and green flares to signal the flight's passage over their airfield, as the bomber would be too high to be seen from the ground. But the plan went awry when Irvine approached earlier than expected at fifteen thousand feet in icy conditions. He didn't know his exact position above a heavy overcast until a radar operator below told him, "Colonel Irvine, I have you. You are three miles north of Northolt."[8]

Fighter Command said later it got only six minutes' warning before the bomber's arrival about 9:45 p.m. "We couldn't get the Mosquitos up in time to meet her," a spokesman said. "The pity about it all is that the good old Dreamboat was doing too well, as it were, and spoiled the program by being too early. We had hoped to do the Dreamboat proud but circumstances were against us."[9] The Air Ministry added that it expected a three-hour warning, but an American liaison officer replied that he had no orders to contact the Brits beforehand. By the time the Mosquitoes reached fifteen thousand feet, the *Dreamboat* was well past London and headed over the English Channel.

"Down at the Air Ministry the ears are a bit red because Fighter Command, with its superlative war-time record for interception of enemy bombers before they crossed the coast, did not get its fighters airborne in time to sight the giant aircraft," an Australian correspondent wrote.[10] The snafu was all the more embarrassing because British hams had chatted with Lieutenant Colonel Shannon nearly an hour earlier over open airways. "The Daily Express asked editorially: 'Why did the British miss the Dreamboat?'" United Press reported. "And the British Broadcasting Corporation rubbed salt into the wound by airing the amateur radio operator's conversation with the Superfortress."[11]

RAF operators saved a bit of face by picking up the *Dreamboat*'s flight summary despite heavy static and reconstructing it from recordings. "Trip very interesting so far," the transcript read. "Three big difficulties: Over Pacific bad weather plus heavy gasoline load; second—Alaskan front a little rugged, got through at 6,000 feet; third—ice cap over Greenland had to climb up to 17,000 feet and 20,000 feet over Iceland which higher than expected."[12] The plane reported everything normal until it encountered the disruptive static. London was socked in when the B-29 flew over, and Irvine and his crew never caught a glimpse of it below.

ASIDE FROM SLIGHT DEVIATIONS FOR WEATHER AND TERrain, the *Dreamboat* largely followed the great circle route from Honolulu to London. Continuing along that line would have taken the bomber across the Mediterranean Sea to North Africa near the border between Algeria and Tunisia. A landing field was ready in Algiers in case the plane couldn't continue to Cairo, but Irvine turned his plane onto a more easterly course toward Egypt at the northeastern corner of the continent.

The *Dreamboat* flew past overcast Paris without catching sight of that city either, and by 10:35 p.m., it was fifty miles east of the capital. French controllers at Orly Field warned the crew of a low-pressure system building over the Adriatic Sea. They also advised Irvine to swing south of his planned course to avoid thunderstorms. As it passed over Geneva, Switzerland, the plane radioed back to Paris,

"Everything is smooth."[13] By 11:30 p.m., the bomber was beyond the towering Alps and approaching Turin in northwestern Italy. The pilots shut down one engine and feathered its prop to reduce speed and conserve fuel, the same tactic Irving used a year earlier during the Guam-to-Washington flight. Weariness was setting in, and unlike the *Turtle*'s crew, the *Dreamboat*'s fliers used the stimulants issued to them. Major Brothers said later they "took booster treatment to fight off fatigue, and after 32 hours of flying the men took a Benzedrine tablet to keep awake."[14]

The *Dreamboat* continued down the length of Italy's boot at fifteen thousand feet. The rugged terrain below might have triggered memories for navigators Hays and Brothers, both having flown there during the war. The bomber passed Rome in the dark and made radio contact with the airport at Foggia, near the east coast and down toward the country's heel. The B-29 then left Italian airspace, continued over the Ionian Sea, and crossed a bit of neighboring Greece. At 3:20 a.m. Greenwich mean time, the *Dreamboat* was flying over the Mediterranean island of Crete.

Major Hays climbed into the nose to watch for thunderheads as the bomber flew into storms before dawn and lurched through turbulence, lightning, and clear ice. Remarkably, the *Dreamboat* also encountered Saint Elmo's fire, the same phenomenon that startled the *Turtle*'s crew over Utah. Light bounced off the nose and around the plane, and it blocked radio reception with static. Major Hays thought the fiery display topped even the northern lights. "The best aurora we saw was in the Mediterranean when electrical discharges caused corona effects around the plane's nose," he said.[15] The weather cleared shortly before sunrise, and the *Dreamboat* flew serenely again on toward the desert, the sky around it filling with white clouds and light.

LISTENERS HEARD NOTHING FROM THE BOMBER AS IT FLEW low over the sea separating the continents. "The almost constant communication which the *Dreamboat*'s radio operator had kept with the ground while over England, France and Switzerland was broken in the last hours of the flight."[16] But there was little anxiety over the crew's safety despite its thirty-six hours in the air. The bomber's destination was John H. Payne Field, "Africa's LaGuardia Airport," according to the War Department.[17] The AAF had built the huge airfield outside Cairo during the war and soon would turn it over to the Egyptian government. The B-29 contacted the tower at 7:40 a.m. local time on Sunday, October 6, two hours ahead of Greenwich mean time and twelve and a half ahead of Hawaiian standard time. The plane "kept up a steady stream of messages, giving its location as it approached the African shoreline." One message fulfilled promises made in Hawaii: "Please inform (King) Farouk we bring special edition Honolulu Advertiser specially printed in his honor. Also bringing special message for United Press. Reserve accommodations Shepheards."[18]

The *Dreamboat* now was running seriously low on fuel. Irvine and his navigators had given up any idea of reaching Khartoum or Wadi Haifa and were beginning to wonder whether they would even get to Cairo. The colonel said later he could have reached Khartoum, a thousand miles past the Egyptian capital, if he'd been able to fly the whole way at the plane's most efficient altitude of ten thousand feet; "however, the crew was 'playing for keeps' and the Dreamboat had to climb many times over ice-forming clouds, he added."[19] Irvine grew more worried as the *Dreamboat* made landfall at Alexandria northwest of Cairo. "We request permission to make a straight in landing instead of a traffic pattern," he radioed. "We are low on petrol and the crew is very tired. We estimate that we have less than 400 gallons of petrol, barely enough to wet the bottom of our tanks."[20]

Payne Field asked if he wanted to fly past the pyramids for photographs, but Irvine was much too low on gas to oblige. "I declare emergency," he radioed, fearing the little fuel still aboard might unbalance the B-29 on landing.[21] He asked for ground crews to clear the field of aircraft and obstacles. With all four propellers again spinning, Irvine brought the plane in with its nose nearly parallel the ground, another maneuver he had used in Washington at the end of the Guam flight to keep the last of his gasoline in the forward tanks. "The giant tires of the super-fortress Pacusan Dreamboat smacked down on a concrete runway of Payne field at 9:45 a.m. . . . and out climbed 10 tired, bearded men who had proved in a 9,500 mile, non-stop flight that the roof of the world is an aerial express highway," United Press reported.[22] The *Dreamboat* had been in the air slightly over thirty-nine and a half hours.

Irvine was first to pop out of the plane. About fifty people waited to greet him in the hot Egyptian sunshine, including military and civilian officials, representatives from the Aero Club of Egypt, and a gaggle of journalists. The colonel shoved an unlit cigar into this mouth as his rumpled and exhausted crew tumbled out onto the concrete apron behind him. Most were sleepless during the flight, Irvine said, but "being the old fud in charge, I got a couple of hours."[23] As the others answered questions, the colonel stepped away briefly with Master Sergeant Fish to inspect the *Dreamboat.*

"Fish, what do you think of the plane?" Irvine asked. His able young crew chief from Appleton made a quick check and said, "All we need is some gasoline and we can take off."[24]

The crew climbed onto a bus and rode to an officers' coffee shop. Like the *Turtle* fliers in Columbus, the *Dreamboat*'s crew also got a quick medical check. "They had two doctors to meet us when we landed," Brothers recalled with a laugh, "but we were in such good condition the doctors left in disgust."[25] Then the ten Americans went along to their Cairo hotel as the press services flashed bulletins announcing their landing to the United States. Anxious relatives finally relaxed, with the Irvine family receiving the news in the pre-dawn hours at St. Paul.

"Good," George Irvine said, "now I can go to bed and get some sleep."[26] Beatrice Warren had tried to keep calm overnight in eastern Nebraska, telling a reporter her husband expected to get a furlough after the flight. "We're going to stay right in Omaha," she'd said, "and play golf—lots of golf."[27] She huddled over a teletype machine for almost six hours waiting for news. "Thank goodness," she breathed when it came. "I'm glad it's over. I knew they wouldn't take any chances but—."[28] Later she showed her two-year-old son a wire photo of the B-29 setting safely in Egypt. "Look, Johnny, there's old Bev," she told the boy.

Staffers at the *Honolulu Advertiser* sighed in relief as well. Buck Buchwach remembered the Egyptian special edition and fired off a cheeky cablegram to Colonel Irvine. "Thanks millions for delivery of papers," it read. "The Advertiser and the Army Air Forces are very proud of prompt and most capable newsboys. Aloha."[29] Major Hays later sent two airmail letters back to the *Advertiser* as souvenirs, each with a copy of the papers flown aboard the *Dreamboat*.

King Farouk also replied to the newspaper days afterward in cablese via United Press correspondent Walter Collins: "EYE RECEIVED COPIES HONOLULU ADVERTISER PROWHICH MANY THANKS STOP EYE SEND MY GREETINGS TO PEOPLE HAWAII AND SEND MY BEST WISHES CREW DREAMBOAT WHO EYE HOPE WILL GOOD VOYAGE HOME."[30]

Apparently even some in the AAF had difficulty accepting the magnitude of what Bill Irvine and his *Dreamboat* crew had accomplished. "When the pilot filed his report for this flight the record section rejected it three times as incomplete," an Oklahoma City newspaper recalled years later. "Only one take-off and one landing were listed. By the standards of the day there should have been at least two, probably three, of each."[31]

THE AAF HAD OFTEN STATED THE DISTANCE OF THE Honolulu-Cairo flight as 10,000 miles, with 10,300 miles popping into print as well. Both figures were high. Confusion over statute miles and longer nautical miles further muddied the reporting and embarrassed the army, although it was not through any error by the B-29 crew. "Maj. N. P. Hays . . . said today the distance covered by the flight over the top of the world was about 9,500 statute miles, instead of 9,500 nautical miles or about 10,900 statute miles, as announced yesterday," the AP reported.[32] An aviation columnist gleefully added, "AAF officials admitted the error and said they weren't trying to belittle the *Turtle*. Unification, boys, it's wonderful!"[33]

Neither figure approached the *Truculent Turtle*'s new record. The Fédération Aéronautique Internationale later calculated the *Dreamboat*'s flight as 9,422 statute miles via three connecting great circle routes: Honolulu to Reykjavík, Reykjavík to London, and London to Cairo. The Superfortress, like the Neptune before it, undoubtedly flew somewhat farther. "The only new record chalked up by the Pacusan Dreamboat . . . was its average speed of 240 miles per hour for a flight of great distance, Col. C. S. (Bill) Irvine, commander, said."[34] But the epic hop over the North Pole captured American imaginations, exactly as Irvine and the AAF had hoped. The army's transpolar hop generated at least as much good press as the navy's jump across the Pacific and possibly more.

Aviation writer and pilot Gill Robb Wilson had monitored the Superfort's radio traffic from the Faroe Islands all the way down to the Mediterranean. He counted himself among the *Dreamboat*'s many admirers but praised the fliers more than the aircraft, marveling at how they had meticulously planned and executed the transpolar flight. "The only super-special equipment with which this plane was fitted for her Honolulu-to-Cairo flight was her crew," he wrote. "They are the dream element in the boat. . . . The Dreamboat is a good big airplane but never any better than the men who use it to accomplish a specific purpose."[35]

SIXTEEN

Blue Skies

TOM DAVIES, HIS *TRUCULENT TURTLE* CREWMATES, AND their wives met President Truman at the White House on October 4, twenty-five minutes after the *Pacusan Dreamboat* took off from Honolulu for Cairo. The dapper chief executive posed for photos in the Rose Garden with the fliers and a flotilla of admirals, and grinned through a five-way handshake with the crew alone. The president was eager to hear about Joey. Davies explained that the little kangaroo was a bit too scrappy to visit the most famous home in America. "'However, that isn't the main reason we didn't bring her with us,' Davies continued. 'You see, Joey isn't housebroken yet, and we figured that the White House wasn't a good place to bring her under those conditions.' Truman said he quite understood."[1]

The president later replied to the letter Davies carried to America from Lord Mayor Totterdell of Perth. "I am confident that the many happy associations which were established during the war will continue and be strengthened during the coming years as new developments in transportation bring our two peoples increasingly closer together," Truman wrote to Australia the following month. And although the *Turtle* hadn't landed in the Northwest as originally announced, Seattle mayor W. F. Devin likewise replied to Totterdell, extending "to you and the citizens of Perth warm greetings from this beautiful city."[2]

The naval aviators got even busier after their White House visit. "The Turtle has become so famous that we have been asked to appear at several air shows and at other places," Rankin wrote to his parents.[3] The navy scheduled the plane for events on both coasts and at several cities in between. Davies meanwhile ribbed his crew's junior member, Roy Tabeling, who'd been stuck on a postwar waiting list for a new car. Shortly after arriving from Perth, the lieutenant commander "received a wire from a Detroit dealer . . . stating that a Buick was being held for him," columnist Drew Pearson reported. "Groans Tabeling's boss, Comdr. Davies: 'I'm in the market for a Chrysler, but so far our flight has inspired no dealer to crash through.'"[4]

Davies and Reid stayed in Washington while Rankin and Tabeling flew the *Turtle* to its first appearances in California. Tabeling's luck deserted him in Oakland, where he ruptured an eardrum on landing and was briefly hospitalized. Rankin and a relief pilot flew the Neptune down to its home in Burbank, where Lockheed engineer Mac Short, now a rising vice president, greeted the plane at the airport. A *Los Angeles Times* reporter admired the *Turtle*'s condition after its record-setting flight. "There's not a scratch or a scar on her. And inside she's as trim as a yacht except for the huge extra gas tanks that almost choke her fuselage amidships."[5]

Rankin next went home to Sapulpa, Oklahoma, where civic leaders handed him the key to the city during an open house held by his parents. The commander hoped to swoop over his hometown in the *Turtle* to entertain schoolkids when he left Tulsa the next morning—Bill Irvine had often done the same thing at St. Paul—but the weather dashed Rankin's scheme. He later flew the *Turtle* southwest to Oklahoma City for a meeting of the National Aviation Clinic, where he delivered a talk the navy had prepared for him titled "Narrowing Horizons."

The commander spoke to a thousand people at a luncheon, making what a local newspaper called a "veiled plea for retention of a separate navy air arm." Rankin disagreed with a speaker who said unification under a single defense secretary would minimize service rivalries and duplication. The fighter plane, the *Turtle* flier said, was "the only combat airplane the army air forces and the navy have that is common to either service." But even the fighters differed significantly, Rankin added, since navy planes were built heavier and sturdier to withstand the rigors of carrier landings. "He called on clinic representatives to express themselves on 'what you honestly think today' in order to 'avoid tragedy tomorrow.'"[6]

Rankin also joked at the luncheon that the *Turtle* was meanwhile "hobnobbing with some B-29s" out at Tinker Field.[7] The pilot did some fancy hobnobbing himself later, taking Gen. Jimmy Doolittle and a visiting New Zealand army officer up for a hop in the famous blue Neptune.

THE *PACUSAN DREAMBOAT* RETURNED TO THE UNITED STATES from Cairo after stops in Wiesbaden, Germany, and Paris. Colonel Irvine flew the first leg in perfect weather to the big American air base near Frankfurt, where someone asked the crew about the plane's name. "I guess that was for our dream of the record," Frank Shannon wistfully replied. "Maybe we'll get another shot at that dream."[8] News agencies soon began speculating about the B-29 flying over the South Pole as well: "It was understood that the flight may be made between Karachi, India, and Hawaii."[9] Papers also speculated that Irvine might lead a formation of B-29s on a flight around the world, and he conceded it had been "on our minds for some time."[10] But the U.S. State Department didn't think much of such sabre rattling, and neither flight ever happened.

Irvine wanted to set a transatlantic speed record flying from Paris to New York and showcase the speed the *Dreamboat* hadn't been able to display while conserving fuel on the Cairo hop. But he turned the plane back after three hours because of fouled spark plugs, surely frustrating crew chief Master Sergeant Fish. The AAF then ordered Irvine to bypass New York for a conference at Westover Field in Massachusetts. The *Dreamboat*'s flight to New England was uneventful except for a brief electrical storm—again three hours out of Paris—that knocked out the radio. The B-29 touched down a few minutes past four o'clock in the afternoon on October 16 after more than twelve hours in the air. "I made no attempt to make a record," Irvine said. "I wouldn't be interested in a record to Westover."[11] Some of his crew suffered from dysentery they had picked up in Egypt, and Fish briefly settled into the base hospital. He kept a careful eye there on an unopened "prized bottle of a three star cognac brandy 'millions of years old'" that an old friend from the Wright company had given to him in Cairo.[12]

The next day Irvine and his crew flew their bomber down to Bolling Field outside Washington to a greeting from generals, an army band, military police standing at attention, and a crowd of well-wishers. "They emerged from the *Dreamboat* as from a bandbox—uniforms pressed, shoes shined, all freshly shaven, to receive warm handshakes and decorations, ranging from the Distinguished Flying Cross to the Air Medal, from Gen. Carl A. Spaatz, Commanding General of the Army Air Forces," the *New York Times* reported. Spaatz also awarded the crew a citation for "meritorious and extraordinary achievement."[13] Even better, photographers captured Frank Shannon getting a passionate welcome-home kiss from his fiancée, Elsie Patzit of Philadelphia. Capt. Ruth Saltzman was more circumspect but posed beaming for a photo with Bill Irvine and General Spaatz.

Reporters clustered around to ask Irvine about the possibilities of flying over the Arctic. The *Dreamboat*'s commander predicted that northern flights would become "routine for all types of airplanes."[14] He added that during the war, General Whitehead had loaded bulldozers onto C-47s and crash-landed them into jungles, where the army then built airfields. "If we wanted to establish a base in the Arctic we could use the same tactics we used in the Pacific," Irvine said. "We could crash land a transport plane or two loaded with bulldozers on a smooth place on the ice. The bulldozers could then smooth out a field in short order."[15]

Wearing crisp khakis, Irvine and his crew visited the White House the following day. President Truman told the airmen they had done "a fine job for our country" and gathered with them around a large globe to trace their flight path for photographers.[16] Major Brothers noticed that Truman was interested in aviation and kept pictures of airplanes in his office. Everyone got a laugh, he said, about the president congratulating them on the performance of a B-29 without anyone from Boeing in the room. (Civilian Richard Snodgrass from the company didn't accompany the army fliers during the visit.)

Brothers hopped a plane to Dayton that evening to see his wife and meet their infant son, Ted. Bev Warren took the *Dreamboat* on what was probably its shortest flight ever, nine miles in three minutes from Bolling Field to Andrews Field in suburban Maryland, where the B-29 was thoroughly cleaned before starting a cross-country publicity tour. On October 23, the crew flew to Southwest Airport in Philadelphia, where hometown hero Frank Shannon signed autographs for his new fans. Two days later, the *Dreamboat* roared north to New York's LaGuardia Field. Major Hays was delayed in Washington, but Captain Saltzman, "who is assisting Col. Irvine on his official report of the polar flight," once more accompanied the crew.[17] Boeing's Snodgrass went along too this time. A twin-engine Beech C-45 Expeditor, the army's version of the popular Beechcraft M-18 civilian light transport, circled the field beside the arriving bomber. The contrast between the two planes emphasized the B-29's great size to those on the ground.

New York's acting mayor (standing in for the vacationing chief executive), the city's police commissioner, and an army general greeted Irvine on landing. The afternoon was wildly festive and hectic as a hundred thousand New Yorkers greeted the *Dreamboat* crew with a parade up lower Broadway. It was bigger than the postwar procession Brothers had enjoyed with General Clark in Chicago, and ticker tape floated down from open windows throughout Manhattan's financial district. An AAF band and mounted police preceded the nine-car motorcade escorted by fifteen motorcycle cops. Afterward fifteen thousand New Yorkers assembled for ceremonies at City Hall, where Irvine laid out the purpose of the AAF's long-distance flights. "Our mission," the colonel declared, "is to go anywhere in the world any time we want."[18] The AAF might have lost the race for a new nonstop distance record, but it was soundly thumping the navy in the public relations battle.

Irvine and his army fliers then enjoyed a luncheon at the Exchange Club hosted by the New York Stock Exchange. Next came a Pan Am cocktail party in the Wings Club at the Biltmore Hotel. It was all heady and unforgettable. "Mutt" Irvine, like the great bedeviled baseball pitcher Grover Cleveland Alexander, had come a very long way from sleepy St. Paul, Nebraska.

SEVENTEEN

Commercial Air

AVIATION EXPERTS FIRST SPECULATED ABOUT AIRWAYS over the Arctic soon after World War I, although dirigibles figured into their predictions as much as the flimsy biplanes of the day. A *National Geographic* writer in 1922 compared the frozen Arctic Ocean to an impassable Mediterranean Sea separating continents. "In the near future it will not only become passable," he wrote, "but will become a favorite air route between the continents, at least at certain seasons—safer, more comfortable, and consisting of much shorter 'hops' than any other air route that lies across the oceans that separate the present-day centers of population."[1]

The AAF and the navy had no doubts in 1946 about what the hops of the *Truculent Turtle* and the *Pacusan Dreamboat* meant for the United States and for the world. Immensely pleased with the Neptune's performance, the navy placed a large order with Lockheed the day after the *Turtle* landed in Columbus. "Navy officers said the planes would be formed into squadrons and placed at strategic naval bases throughout the world as soon as deliveries were made," the United Press reported. "They did not reveal the size of the contract. . . . As designed for tactical operations for the navy, the P2V planes bristle with offensive weapons."[2]

As for the army, General Spaatz called the *Dreamboat*'s transpolar flight "an epochal achievement in aeronautical history. It gained valuable information on navigational, engineering, communications, weather, fuel consumption and physical endurance problems. It proves the feasibility of a flight across the polar wastes by properly equipped aircraft."[3] The *Philadelphia Inquirer* noted in an editorial, "General Spaatz, diplomatically, did not refer to the war possibilities of Polar flying, but some unnamed American military experts remarked that the big B-29's air journey more than proves the contention that this country is vulnerable to air attack across the vast ice cap at the top of the world."[4]

Public affairs officers didn't need to bang their drums too strenuously to relay Spaatz's strategically important message to civilians still weary from the last war. "We played a game today with a piece of string and a globe," an editorialist in Milwaukee wrote immediately after the *Turtle* flight. The writer stretched the string from Perth to Columbus, snipped it at both ends, then ran the long strand here and there across the globe to see how many places it reached from Ohio . . . and it was almost everywhere. "And the inevitable thought followed—what one airplane can do, others can do. What an American plane can do, planes of other nations can do. Try the game of the string and the globe."[5] Many Americans did exactly that and reached the same worrisome conclusion. "The historical flight of the Truculent Turtle should be remembered by every American when he counts his future, for it wrote the official end to our ocean barriers as a factor in international politics or war," the *New York Post* said.[6]

The flight of the bigger, heavier, and more lethal *Dreamboat* soon afterward reiterated the message and drove the shaft deeper. "It should be clear now that mechanical wings can carry us or anybody from any point on earth to any other in a single burst of speed," the *New York Times* said. "If a general war is ever again allowed to scourge mankind there will be no islands of safety anywhere."[7] PM also noted in New York City, "The Pacusan Dreamboat is actually obsolescent.

Other bombers dwarfing the B-29 are on drawing boards or actually airborne."[8]

A small newspaper in upstate Rome, New York, thought the armed forces had made their point only too clearly. "Indeed, there is no obvious reason for similarly long non-stop flights in the armed services," its editorial observed. "Presumably the Turtle and the Dreamboat tests come mainly within the field of propaganda."[9] But a paper in Elmira added, "Viewed from a more cheerful angle, the Truculent Turtle's record flight shows the increasing capacity of aircraft to give useful service to mankind."[10]

The flights of the *Turtle* and the *Dreamboat* also offered insights into the future of global commercial aviation. "What military aircraft perform today, transport planes duplicate tomorrow," the Associated Press observed in the spring.[11] The implications of the AAF's Honolulu-to-Cairo flight were particularly important, with the *New York Times* noting, "The top of the world has become an air route with vast possibilities." The *Times* cautioned that the Arctic needed more communications and weather stations, as well as emergency landing fields, before flights linking America, Europe, and Asia became commonplace. "All these facilities will come in time and it is possible that at a not-too-distant date commercial air lines will have regular services across the polar wastes."[12]

Bill Irvine concurred. "The trans-Polar route represents to modern military and commercial airplanes what tunnels through mountains once represented to railroads," he wrote soon after his Cairo hop. "A new age has opened in the story of swift communications between continents and hemispheres."[13] The navy magazine *All Hands* expressed a similar view. "The world looked even smaller when the public realized these [two military] flights had been made by equipment at hand, not fanciful eight-jet propelled bombers of the future," it observed.[14] Civilians soon recognized the portents and importance of the flights too. "Commercially, the seven-league jumps which British and American planes have been attempting since the war's end have striking possibilities," commented a newspaper in Lincoln, Nebraska. "Wherever time is a factor, long-distance flights will be invaluable for linking worldwide markets, raw material sources, prompt checkups on scattered local situations and the like."[15]

Irvine was still in Egypt when he spoke about the possibilities of new, shorter airways between Asia and Europe. "It would appear to me that the shortest route—from the commercial viewpoint—from London to Japan and other Far Eastern points is over the polar region," the colonel said. "It should take 28 hours from London to Japan in a commercial version of the B-29." Flying over the Arctic, he added, was "both safe and practical."[16] Irvine observed later that the flight also had "proved inter-plane and ground communications could be conducted in arctic regions."[17] Frank Shannon agreed. "If commercial airways decide to use this region as an air route," the affable lieutenant colonel said, "I believe they could have successful communications with a few well placed low frequency radio stations."[18]

The *Pacusan Dreamboat* made its flight to Cairo using only standard navigation gear. "But we know now that [these instruments] are too small, light and inaccurate for routine operations across the pole," Irvine said in America. The AAF consequently recommended changes. "With those changes, Polar flights are not only possible but practical for commercial as well as military airplanes," Irvine said. "They will be routine for average type airplanes in the future."[19]

While the *Dreamboat*'s effect on the future of commercial aviation was undeniable, the *Truculent Turtle*'s influence was small or nonexistent, except perhaps to broaden airline executives' thinking about long-distance transport and travel. The Neptune, after all, was a relatively small patrol bomber with no real capacity for carrying freight or passengers. The *New York Times* dismissed the significance of the *Turtle*'s extraordinarily long flight as "almost completely military. Airline people see no immediate or practicable translation of such range into commercial transport of passengers or cargo."[20]

But B-29s such as the *Dreamboat* already were influencing peacetime civilian markets. Boeing soon introduced a luxurious passenger version of the Superfort called the Boeing 377 Stratocruiser, which had a pressurized double fuselage and a staircase between the decks. The model never really caught on, but the company still built fifty-six of the rather chubby-looking airliners between 1947 and 1950.

The humble C-54/R5D Skymaster exerted the greatest influence of all the aircraft involved in the long-distance military flights of 1946. United and Western Airlines already were flying the unpressurized civilian version, the DC-4. Skymasters later would play a significant international role during the Berlin Airlift of 1948–49, ferrying millions of tons of food and cargo to break the Soviet blockade of the isolated German city. Douglas Aircraft introduced the follow-on DC-6 in 1947 in response to Trans World Airline's triple-tailed Lockheed Constellation (and eventually the elegant Super Constellation). The upgraded DC-6B model proved popular with airlines and passengers alike. "Perhaps the epitome of piston-engine airliner design, the Douglas DC-6B combined unrivalled operating efficiency and reliability," reports the Smithsonian National Air and Space Museum. "Its slightly stretched fuselage could carry 88 passengers. DC-6BS entered service with United in 1952, and Pan Am used them to pioneer tourist fares across the Atlantic."[21]

Scandinavian Airlines System (SAS) was the first commercial carrier to offer flights from Europe to Asia via the Arctic. SAS flew DC-6BS from Oslo to Tokyo in May 1953, with an overnight stop in Anchorage, thereby cutting three thousand miles off the conventional route. A year and a half later, in November 1954, the airline began flying the northern route between Europe and the American West Coast, with brief stops for fuel in Canada and Greenland. The *New York Times* called SAS's service "one of aviation's greatest advances." The article's author had flown in the China-Burma-India theater over the Himalayas and the infamous "Hump" during the war. The former AAF pilot noted, "The new polar flights mean that the commercial airliners, like the Air Force's strategic warplanes, are now able to fly to almost all parts of the earth at will. They no longer have to follow aerial cow paths, planning stops at already established refueling stations."[22]

The Collins Radio Company placed an ad in the *Los Angeles Times* congratulating SAS for "opening a new avenue of trade and commerce."[23] Other international airlines weren't far behind in offering transpolar service between Europe and Asia, Hawaii, and the West Coast. Pan Am even used the old string-and-globe concept to illustrate a 1957 newspaper ad urging travelers to "take the 'short cut' to Europe" via the Arctic from Seattle, San Francisco, and Los Angeles.[24]

Engines, instruments, and airframes all steadily improved, often propelled by military innovations. The introduction of the now-iconic Boeing 707 in late 1957 ushered in the jet age of passenger air travel. But at mid-century, as today, polar flights nearly always followed the northern routes rather than the southern ones, if only because of geography. "Lack of airports for diverting in case of emergency and security restrictions keep scheduled flights from flying over the South Pole or, indeed, the 'White Continent' as a whole," a travel magazine explains.[25] A sight-seeing flight to Antarctica ended in tragedy in 1979 when an Air New Zealand plane crashed into Mount Erebus during a whiteout, killing all 257 people on board. The airline never offered such flights again, although a few charter services do. And while no airline currently flies a scheduled route over the southern pole, some do fly near Antarctica—on flights from Sydney to Johannesburg, for instance, or from Auckland to Buenos Aires.

Other tragedies following Mount Erebus also highlighted the limitations of the Arctic airways at the top of the world. The *Turtle* and *Dreamboat* crews flew routes that avoided Soviet airspace in 1946, restrictions that remained in place for several decades. In April 1978, a Soviet warplane fired on Korean Air Lines (KAL) Flight 902 flying a polar route from Paris to Seoul when the 707 accidentally strayed over Soviet airspace near Murmansk. With two passengers dead and ten wounded, the crew managed to crash-land the crippled plane on a frozen lake. KAL later said the jetliner had gone a thousand miles off its intended course "due to a defect of the directional gyro."[26] The Soviets treated the survivors well once the mistake became apparent.

The next international incident was far worse. A Soviet missile downed KAL Flight 007 September 1, 1983, killing all 269 people on board, after the Boeing 747 jumbo jet also wandered off course during a flight from New York to Seoul via Anchorage. Russian officers apparently mistook the airliner for a U.S. military intelligence-gathering plane flying nearby in international airspace. Amid escalating international tensions, President Ronald Reagan denounced the shootdown as "a horrifying act of violence."[27] Russia finally opened air routes over its far eastern territories in 1993 after the Soviet Union's dissolution. Following extensive negotiations, the country allowed new air routes over the Arctic in 1998–99. "Of course, that was a bitter lesson," a Russian aviation official said of the KAL 007 disaster. "A terrible thing. I can give a 100 percent guarantee that won't happen again."[28]

No one guaranteed that the Russian routes will always remain open, however, as an international incident perhaps portended in May 2021. "The airspace over Eastern Europe turned into a geopolitical checkerboard on Thursday," the *New York Times* reported, "as Russia rejected some European flights that were avoiding Belarus, the latest salvo in the furor over the forced landing of a passenger jet with a Belarusian dissident onboard."[29]

United Airlines Flight 895 from Chicago to Hong Kong meanwhile became the longest scheduled passenger flight in the world in 1996. A slightly longer New York–to–Hong Kong flight over the north saved over four thousand gallons of fuel and two hours in the air compared with a conventional route. (Airliners flying from Asia to the United States or Europe, in contrast, generally benefit from strong tailwinds by following more southerly routes.)

"The north is kind of a last frontier," an official at NAV CANADA, which oversees the country's civil air-navigation services, said in 2000. "It's as exciting as when the Pacific opened up." Flying in the Arctic still remained tricky, with compasses being unreliable and geostationary navigation satellites rendered useless "because they hang over the Equator and are not visible from the pole," the *New*

York Times explained. "But the Global Positioning System works, and so does inertial navigation."[30]

The satellite-based Future Air Navigation System and the Airborne Communications Addressing and Reporting System (commonly known as ACARS), together with other new or improved technologies, further advanced safe navigation near the geographic and magnetic North Poles. By 2010 fifty thousand flights annually crossed the Arctic, connecting Europe, North America, and Asia. "Today, uncounted passengers and tons of cargo fly across the North Pole each year," an Alaska newspaper noted in 2013. Recalling the pioneering Soviet pilot who disappeared while flying in the Arctic seventy-six years earlier, the writer added, "Levanevsky's vision of commerce and communication flying over the Arctic Ocean has come to pass, as he seemed confident it would."[31]

AS WITH MANY SCIENTIFIC OR TECHNOLOGICAL ADVANCES, increased northern air traffic also brought unintended consequences. Natural radiation entering Earth's atmosphere from space registers at a very low level on the ground, but it increases with altitude—such as when flying in a jetliner—and climbs considerably higher the nearer you get to the poles. When the National Aeronautics and Space Administration (NASA) studied polar flights during a solar storm in 2003, the results "showed that passengers received about 12 percent of the annual radiation limit recommended by the International Committee on Radiological Protection. The exposures were greater than on typical flights at lower latitudes, and confirmed concerns about commercial flights using polar routes." The researcher added that while one or two flights wouldn't expose passengers or crew to high doses of radiation, "multiple flights definitely pose an additional health risk."[32]

Another issue arose simultaneously with concerns about global warming and dwindling Arctic sea ice. "A new study suggests one way that humans could slow the melting of the sea's ice—by preventing international flights from crossing over the Arctic circle," the

Washington Post reported in 2012.[33] The climate researchers who conducted the study wrote that transpolar flights were persistent sources of black carbon and other pollutants that contribute to the ice melt in the Arctic. But there is a trade-off to abandoning the northern routes: planes must fly farther to reach their destinations, causing problems elsewhere farther south. "Although the slight CO_2 emission increase due to rerouting may dampen the benefit of rerouting over many more decades," the authors concluded, "rerouting or partial rerouting (allowing cross-polar flights during polar night only) should delay the elimination of Arctic sea ice, which is otherwise expected in 2–3 decades."[34]

Contrails, the distinctive white condensation trails left behind by high-flying aircraft, are another growing concern. Climatologists suggest that the trails' heat-trapping qualities might contribute to global warming. At least along Arctic air routes, however, contrails seem to have little adverse effect. "Flights close to polar regions tend to produce only thin, hazy contrails," NBC News explains, "because the dry air there lacks sufficient moisture to form significant ice crystals."[35]

Paradoxically, the terrible drawback of the Arctic airways is also their greatest benefit. "The *Dreamboat* was pioneering a route that countless thousands in this Air Age will follow," a Boy Scout magazine said in early 1947, "for with the coming of the Air Age man will no longer be bound by geographical limitations in his travels."[36] What was true of piston-driven aircraft capable of delivering either atomic weapons or peaceful passengers during the mid-twentieth century remains true for fly-by-wire airliners, stealth bombers, and hypersonic intercontinental ballistic missiles. And proving the viability of such limitless airways was always the primary mission of both the *Turtle* and the *Dreamboat*.

Epilogue

IT WAS HARDLY SURPRISING WHEN THE *TRUCULENT TURtle* and *Pacusan Dreamboat* crossed paths soon after completing their historic flights. Both planes flew into Cleveland for the National Aircraft Show in mid-November 1946. The show site was a former aircraft plant beside the city's airport. "The bomber plant, turned into a huge display case for the aviation industry, is one of the few buildings in the world large enough to house the tremendous assembly of planes and equipment," the United Press reported.[1] Hometown pilot Commander Davies was the featured guest one day, Colonel Irvine and his crew the next. Visitors inspected both planes during the ten-day event.

A month later, newspapers reported that the *Turtle* was "being groomed for a flight to the South Pole with Adm. Richard E. Byrd as the passenger. . . . A major purpose of the flight is to test the distance champion of the air in Arctic flying weather as it joins the navy's expedition to Antarctica or Little America."[2] In mid-January 1947, the *New York Times* reported that Davies would fly the *Turtle* to Little America with Rankin accompanying him in a second Neptune. "Both patrol bombers will carry scientific equipment, including radar and three-dimensional cameras."[3]

The *Turtle* never flew to Antarctica, but Davies was assigned to Rear Admiral Byrd's staff. "While serving in that post he invented the sky compass, a device for navigating near the Earth's magnetic poles."[4] Rankin also received orders for temporary duties connected with the admiral's South Pole expedition, dubbed Operation High Jump. The other members of the *Turtle* and *Dreamboat* crews likewise continued their military careers, with several later finding success in business as well.

BILL IRVINE (1897–1975) KNEW HIS TIME IN THE AIR WAS nearly over after his transpolar flight in 1946. "I hate going back to a desk in Tokyo, but there is no escape from the desk-work," the colonel said in Cairo. "It is a necessary thing if I want to prepare for the next job." The globe had been fairly well covered, a correspondent said, "so what about a rocket trip to the moon?" The Nebraskan shot back, "Why not? That's more like it."[5]

The colonel married Capt. Ruth Saltzman (1910–73) in December 1946. Saltzman soon left the AAF while Irvine remained and saw the creation of the U.S. Air Force in 1947. After receiving his long-delayed star of a brigadier general in 1949, Irvine commanded bomber wings in New Mexico and Texas; then he returned to Dayton as the deputy commander for production of the Air Materiel Command. Irvine later was heavily involved in the Boeing B-47 Stratojet and B-52 Stratofortress bomber programs. He earned two more stars before retiring in 1959 as a lieutenant general and deputy chief of staff for matériel at air force headquarters. In retirement Irvine was an assistant to the president in charge of West Coast operations for the Avco Corporation and later served as a consultant for Rockwell International. He remarried after Ruth's death and shortly before his own passing.

"Col. Irvine was a burly, sarcastic officer, noted for getting things done when others thought they were impossible," his old hometown Nebraska newspaper noted after his death in Los Angeles. "His career at times was marked with flamboyance."[6] He and Ruth are buried at Arlington National Cemetery.

Bev Warren (1911–83) not only earned back his colonel's eagle but also rose to major general in the new air force. He held several key posts in the Air Materiel Command before retiring in 1960. He also joined Avco and served as the director of missile operations. The company named him vice president and general manager of its Lycoming Division in 1963. Warren also served on the boards of a bank, a hospital, and a chamber of commerce; played golf; and made furniture for his Connecticut home.

Frank Shannon (1899–1983) stayed in the AAF following the *Dreamboat* flight rather than return to civilian life a third time. He earned a promotion to full colonel and during the early 1950s commanded a communications group in Okinawa. Shannon was the special assistant for electronics in the Directorate of Maintenance Engineering at Wright-Patterson Air Force Base when he retired to Florida in 1957. "Pappy" helped Bill Irvine become an enthusiastic ham radio operator.

Norman Hays (1918–65) became a literal poster boy for military navigation when the air force used his history and photo in recruiting ads for new officers. He left active duty in 1956 after sixteen years to become the director of North American Aviation's Autonetics Division, meanwhile rising to colonel in the U.S. Air Force Reserve. At his accidental death from a fall at home on the Fourth of July 1965, he was North American's corporate vice president for field operations. He too is buried at Arlington. The civilian Institute of Navigation established an annual award in his name that it still bestows today.

James Brothers (1920–2005) stayed in the air force, served in various squadrons, and attended the Air Command and Staff School. In 1950 he was caught in a shootout between police officers and escaped convicts on a dark road outside Little Rock, Arkansas. The officer who warned him to "hit the ditch" was shot twice while warning others. "The car behind mine was hit four times by rifle bullets—one in the windshield near the steering wheel," Brothers said, and the incident was nearly as harrowing as flying a B-29 over Tokyo. "We got into the ditch and huddled there 45 minutes in the rain, until the battle stopped and we learned the convicts had escaped."[7] Brothers commanded a special operations squadron in California and an air refueling squadron in Massachusetts before retiring as a colonel.

Bob Kerr (1917–95) left the air force in 1954 to begin a spectacular business career. He too joined Avco and surpassed all his *Dreamboat* contemporaries, whom he may have helped to recruit. He rose to president in 1960, chief executive officer in 1969, and chairman in 1974. "During his 12-year tenure as chief executive of Avco," the *New York Times* reported, "Mr. Kerr broadened the company's focus beyond defense and aerospace, so that it included businesses like motion pictures, life insurance, financial services and land development."[8] Under his watch, Avco bought Embassy Pictures, produced the smash hit *The Graduate,* and changed the studio's name to Avco Embassy. Kerr also bought up credit card company Carte Blanche. "But he paid a price, Kerr told an interviewer in 1979, two years before retiring. 'I'm 61 years old, going on 90,' he said."[9] He died later of heart failure.

James Dale (1920–2003) stayed in uniform as a bomber pilot. His B-29 was among fifty that took off from overseas bases to mark the air force's first birthday with flights to cities across America; his bomber was the first to land. Dale rose to lieutenant colonel, married a local woman while stationed in Germany, and continued living there during retirement.

Gordon Fish (1923–2005) spent five years testing the B-36 bomber, a mammoth long-range plane with four turbojets forward plus six pusher-type radial engines behind the wings—a crew chief's dream. Fish then earned a commission and rose to major before retiring after twenty-seven years' service. He worked another ten years for General Dynamics in Texas.

Gray Vasse (1919–2009) married a Canadian he met in 1948 while flying air force missions from Edmonton. His new family used his first name and called him Ed. Vasse served at bases around the world before retiring to Spokane, Washington, as a chief warrant officer in 1964.

Richard Snodgrass (1919–2009), the only civilian on the *Dreamboat* flight, married in 1949. He retired in 1986 after forty-five years with Boeing.

The last B-29 rolled off the Boeing production line before the Honolulu-to-Cairo flight in 1946, as the AAF and the air force concentrated instead on the B-36. The first all-jet bombers then replaced the B-36 during the 1950s. Superforts, however, again saw action in the Korean War, with the last B-29 squadron remaining in service until 1960.

The *Pacusan Dreamboat*, Superfortress 44–84061, set two closed-course distance records in flights over the United States in 1947. Colonel Irvine was the project officer rather than the pilot. Dispatched to the Strategic Air Command and modified for photo reconnaissance, the *Dreamboat* was twice damaged during flights in 1948, once by a bird strike and then by bad weather. In August 1954 the bomber returned to Tinker Air Force Base in Oklahoma City "for the last time and was reduced to aluminum ingots at the salvage yard."[10]

TOM DAVIES (1914–91) FLEW THE FIRST NEPTUNE OFF AN aircraft carrier in 1948. He was "back in the news with a wallop" the following year, embroiled in the Revolt of the Admirals and the navy's nasty campaign against the B-36 bomber program.[11] "Cmdr. Davies' acknowledgment he spread unverified gossip about the Air Force came against a background of long-smoldering bitterness between the Navy and the Air Force," the Associated Press reported.[12] The incident was embarrassing but didn't harm his career. "The 34-year-old Cleveland pilot may get a few black looks from the chiefs of defense establishment, but it is doubtful whether his fellow navy flyers will feel he sinned unduly," a hometown columnist wrote.[13]

Davies later commanded the fleet oiler USS *Caliente*, Fleet Air Wing Three during the Cuban Missile Crisis, and NAS Norfolk. He graduated with distinction from the National War College and earned a master's degree from George Washington University. Davies retired as a rear admiral in 1973 after forty years in a naval uniform. He later became president of the Navigation Foundation, which develops and promotes celestial navigation. In 1989 he directed a National Geographic Society investigation into allegations that Robert E. Peary had faked his claim of reaching the North Pole in 1909. Davies exonerated the explorer, declaring, "Peary was not a fake or a fraud."[14]

Gene Rankin (1914–2000) switched to flying single-engine planes and commanded a carrier-based fighter squadron in 1947. He reached Antarctica as part of an expedition to the South Pole in 1954. During the early 1960s, Rankin was the commander of the seaplane tender USS *Pine Island* before commanding the aircraft carrier USS *Kearsarge*. The flattop was the recovery ship for Project Mercury astronauts Walter "Wally" Schirra and Gordon "Gordo" Cooper. "We did something Schirra wanted to do—be recovered in the capsule," Rankin recalled. "No one in NASA or on the Eastern Seaboard apparently even thought it could be done."[15] The Oklahoman retired from the navy as a captain in 1967. After the 1991 deaths of his wife, Virginia, and friend Tom Davies, Rankin married Davies's widow, Eloise.

Walt Reid (1914–94) commanded the seaplane tender USS *Currituck* during the early 1950s; then he retired as a captain in 1962. He briefly managed apartment complexes in suburban Washington before working as a management analyst for the U.S. Coast Guard.

Roy Tabeling (1920–94) married in 1948 and began a family while serving at various duty stations, including Ford Island at Pearl Harbor and the Pacific Missile Range at Point Mugu, California. Following retirement as a full commander after twenty years in the navy, Tabeling worked for RCA, a prime contractor for the Atlantic Missile Range. He later happily shed his necktie to open a surf shop in Florida. Following a divorce, he lived across from Patrick Air Force Base near Coco Beach, where he kept an eye on the arriving and departing planes.

Robert Bailey (1913–77), the *Turtle*'s project engineer, continued a steady rise within Lockheed California, organizing the industry's first military operations research organization in 1948. Bailey held increasingly important titles over the following decades, including vice president, chief engineer, chief spacecraft engineer, and vice president for commercial air transports. His work involved programs ranging from the turboprop Lockheed L-188 Electra (the military version, the P-3 Orion, supplanted the aging Neptune) to supersonic transports and orbital laboratories. Bailey retired from Lockheed in 1969 after more than three decades' service and was hailed as "a pioneer in the systems analysis approach to designing new aircraft."[16]

Lockheed P2Vs proved to be rugged, dependable, and long-lived patrol planes. The order the navy placed after the *Turtle*'s globe-girdling flight jump-started over fifteen years of domestic Neptune production. Naval reserve squadrons flew the plane well into the 1970s. Lockheed and its partners built over a thousand Neptunes for the United States and ten other countries including Australia, which saw them first. Manufacturing continued in Japan until 1979, bringing total production to nearly twelve hundred planes.

The *Truculent Turtle*, P2V Neptune BuNo 89082, continued flying for several years. In 1949 it retraced the path flown thirty years earlier by the Curtiss NC-4 flying boat *Lame Duck*, the first navy plane to cross the Atlantic. After its removal from service in 1953, the *Turtle* stood on static display near NAS Norfolk and later outside its main gate. In 1977 the navy sent the plane to the National Naval Aviation Museum in Pensacola, Florida.

A jet-powered U.S. Air Force B-52 bomber broke the *Turtle*'s long-distance nonstop record in 1962 with a flight from Okinawa to Spain. But the Neptune's record for piston-driven flight stood for more than forty years until Dick Rutan and Jeana Yeager flew around the world in December 1986 aboard designer Burt Rutan's lightweight experimental aircraft *Voyager*. "One of the pilots who set the piston-driven flight record, Roy Tabeling, now 66, of Merritt Island, Fla., said he was eagerly tracking the Voyager," the AP reported during the flight.[17]

The restored *Truculent Turtle* is displayed today in Pensacola. Visitors to the aviation museum there see it suspended from the ceiling as if still in flight over the blue Pacific, bound from Australia to America or beyond with four airmen and a young kangaroo aboard.

ACKNOWLEDGMENTS

THANK YOU TO MAJ. GEN. ED MECHENBIER, USAF (RET.), for advice and support; to LynnAnn Tabeling and Kenneth Horner for sharing memories of their fathers; and once again to agent Kelli Christiansen for seeing the appeal of the whole thing.

APPENDIX

Crew Rosters

Truculent Turtle (Perth–Columbus), USN

Cdr. Thomas D. Davies, Cleveland OH, command pilot
Cdr. Eugene P. Rankin, Sapulpa OK, pilot
Cdr. Walter S. Reid, Washington DC, pilot
Lt. Cdr. Roy H. Tabeling Jr., Jacksonville FL, pilot

Pacusan Dreamboat (Honolulu–Cairo), USAAF/Boeing

Col. Clarence S. Irvine, St. Paul NE, command pilot
Lt. Col. Beverly H. Warren, Plainview TX, copilot
Lt. Col. Frank J. Shannon, Philadelphia PA, communications
Maj. Norman P. Hays, Grove OK, navigator
Maj. James T. Brothers, Fountain City TN, navigator
Maj. James R. Kerr, Arcadia CA, engineer
Maj. James R. Dale, Wise VA, third pilot, flight engineer
M.Sgt. Gordon S. Fish, Appleton WI, crew chief
M.Sgt. Edward G. Vasse, Huntsville MO, radio operator
Richard B. Snodgrass, Loma CO, Boeing flight engineer and observer

NOTES

Except in quoted materials, Russian personal names appear with current anglicized spellings. Place names from the 1930s and 1940s appear with current spellings in italics immediately following.

Abbreviations

ENHP Edward N. Horner Papers
RHTP Roy H. Tabeling Papers
USAFB United States Air Force Biographies

1. The Great Circle

1. "Giles Urges Air Force Occupation of Japan 'for Next 100 Years,'" *Washington Star*, September 20, 1945. "LeMay always said that the atomic bombs were superfluous," author Malcolm Gladwell adds. "The real work had already been done." Gladwell, *Bomber Mafia*, 188.

2. W. H. Shippen Jr., "Star Reporter Logs B-29 Bombing Raid over Tokyo in Eye-Witness Account," *Washington Star*, April 29, 1945.

3. Nelson M. Shepard, "3 B-29s Make Westbound Hop, Guam to D.C.," *Washington Star*, October 20, 1945.

4. Nelson M. Shepard, "Four B-29s Complete Record Nonstop Trip from Japan to D.C.," *Washington Star*, November 2, 1945.

5. "Army Will Strip Planes of War Paint for Speed," *New York Times*, December 14, 1943.

6. Whittaker, "Meet the B-29!," 12.

7. "B-29 Sets New Record, Will Continue to Luzon," *Honolulu Star-Bulletin*, December 28, 1945.

8. Jacobsen, "Red-Tailed Beauties," 7.

9. Irvine, "Record Flights," 3.

10. "Flying Secretary," 3.

11. "Guam to Washington, Non-Stop!," *Honolulu Star-Bulletin*, November 20, 1945.

12. W. H. Lawrence, "B-29 Sets Record in Guam Hop, Flying 8,198 Miles to Capital," *New York Times*, November 21, 1945.

13. "Great Circle Gives Shortest Route," *Newcastle Morning Herald and Miners' Advocate*, October 5, 1946.

14. "Guam to Washington, Non-Stop!"

15. "Record-Breaking Plane Arrives at Wright Field," *Dayton Journal*, December 1, 1945.

16. W. H. Shippen Jr., "Superfort Flyers Hoarded Gas in Record Non-Stop Guam Hop," *Washington Star*, November 21, 1945.

17. Lawrence, "B-29 Sets Record."

18. "B-29 Smashes Pacific Record; Flies 8,198 Mi.," *Chicago Tribune*, November 21, 1945.

19. Lawrence, "B-29 Sets Record."

20. "B-29 Flies Nonstop from Guam to D.C. for New Distance Mark," *Washington Star*, November 20, 1945.

21. "Strange Men, Be Patient!," *La Crosse Tribune*, November 21, 1945.

22. "Record-Breaking Plane Arrives."

23. "B-29 Crew Bleary-Eyed after Flight from Guam," *Philadelphia Inquirer*, November 24, 1945.

24. Joseph A. Bors, "Distance Mark Set by B-29," *Pittsburgh Sun-Telegraph*, November 20, 1945.

25. "B-29 Flies from Guam to Capital," *Albany (NY) Times Union*, November 21, 1945.

26. Bors, "Distance Mark Set."

27. "B-29 Sets World Mark with 8198-Mile Hop," *Philadelphia Inquirer*, November 21, 1945.

28. "B-29 Flies from Guam to Capital."

29. "'Almost Here,'" *Washington Star*, November 21, 1945.

2. Unification

1. "Eisenhower and Nimitz Prepare to Take Over Top Posts in Services," *Washington Star*, November 21, 1945.

2. Felix Belair Jr., "Truman Will Urge Army-Navy Merger," *New York Times*, November 21, 1945.

3. Robert Bruskin, "Army-Navy Jealousies Impeded War Overseas, '44 Report Says," *Washington Star*, November 4, 1945.

4. U.S. Senate, *Hearings before the Committee*, 413.

5. U.S. Senate, *Hearings before the Committee*, 415.

6. Truman, "Our Armed Forces."

7. Belair, "Truman Will Urge."

8. McFarland, *Concise History*, 40.

9. "Stimson Recommends Unification of Army and Navy after War," *Washington Star*, April 25, 1944.

10. "Eisenhower Statement on Military Merger," *Washington Star*, November 16, 1944.

11. "Legion Speech by Eisenhower: Keep U.S. Strong," *Chicago Tribune*, November 21, 1945.

12. "Services Take Debate on Military Merger to Nation by Radio," *Washington Star*, December 3, 1945.

13. "General Claims Merger Will Bolster Defenses," *Philadelphia Inquirer*, December 16, 1945.

14. "Admiral Fears Chaos in Merger," *Philadelphia Inquirer*, December 16, 1945.

15. Rosenberg, "American Postwar Air Doctrine," in Hurley and Ehrhart, *Air Power and Warfare*, 247.

16. "'Instruments of Policy,'" editorial, *Waterloo (IA) Daily Courier*, October 10, 1946.

17. McCullough, *Truman*, 476.

18. J. A. Fox, "Truman Message to Congress Proposes Army-Navy Merger along War Department Line," *Washington Star*, December 20, 1945.

19. Trest, "View from the Gallery," 18.

20. "Navy Gags Officers to Prevent Voicing Views on Merger," *Washington Star*, December 20, 1945.

21. Trest, "View from the Gallery," 18.

22. Schratz, "Admirals' Revolt," 65.

23. Chris Mathisen, "Calmness Urged by Eisenhower in B-36 Row," *Washington Star*, October 20, 1949.

24. Anand Toprani in Nelson, "Revolt of the Admirals."

3. Cross Country

1. Item, *St. Paul Phonograph*, September 4, 1919.

2. "St. Paul Man Has Fine Role in 'Wings,'" *St. Paul Phonograph*, November 14, 1928.

3. "Lieut. Clarence S. Irvine Wrecks His Plane to Save Lives," *Manila Daily Bulletin*, reprinted in *St. Paul Phonograph*, March 20, 1929.

4. "Lieut. Irvine Thrills the Local Populace," *St. Paul Phonograph*, March 27, 1935.

5. Pat's Pennings, *Howard County (NE) Herald*, September 6, 1939.

6. "Col. Irvine Is Honored Guest," *St. Paul Phonograph*, December 5, 1945.

7. "Flier Tells Possibilities of Bombing," *Syracuse Herald-Journal*, December 10, 1945.

8. "St. Paul Anxiously Follows Col. Irvine's Record Trips," *Lincoln (NE) Star*, September 7, 1946.

9. "B-29 Flies from Coast in Record 5½ Hours," *Brooklyn Eagle*, December 12, 1945.

10. Royal Riley and Jess Stern, "B-29 Spans U. S. in 5:27," *New York Daily News*, December 12, 1945.

11. "B-29 Flies from Coast."

12. "B-29 Sets Mark in Trans-U.S. Hop," *Philadelphia Inquirer*, December 12, 1945.

13. "B-29 Sets Air Record," *Honolulu Advertiser*, December 12, 1945.

14. "Sees 4-Hour Flight across United States," *Joplin (MO) New Herald*, December 13, 1945.

15. "4-Hour Air Trips to Coast Forecast," *New York Times*, December 13, 1945.

16. "Sees 4-Hour Flight."

17. "4-Hour Hops to Coast 'Only a Year Away,' Says Dreamboat Pilot," *Brooklyn Eagle*, December 13, 1945.

18. Clarence Irvine, "B-29 'Dreamboat' Sets Record for Altitude Flying," *Troy (NY) Record*, December 13, 1945.

19. Julia McCarthy, "Dreamboat's WAC Prefers to Dream," *New York Daily News*, December 13, 1945.

20. Hays, "Across the Top," 28.

4. Neptune

1. "Lockheed Aircraft," *Naval Aviation News*, July 1957.

2. Francillon, *Lockheed Aircraft since 1913*, 261.

3. "American Aircraft in the RAF."

4. "From War to Peace," 11.

5. "Lockheed P2V Neptune," 36.

6. "Plane Designer," *Perth West Australian*, September 23, 1946.

7. "From War to Peace," 12.

8. "Professional Notes: Operation Turtle," 371.

9. "Plane Designer."

10. *Lucky Bag*, 204.

11. Morison, *Atlantic Battle Won*, 220.

12. "P2V Can Carry Atomic Bombs," *New York Sun*, December 19, 1945.

13. Reid, "I Rode the Turtle," 92.

14. "The New Long-Distance Record," *Flight*, October 10, 1946.

15. "Navy's New Super Patrol Plane Tops in Range, Speed and Armament," *Lowell (MA) Sun*, December 19, 1945.

16. "Lockheed's Peace Persuader," *Nashville Tennessean*, December 23, 1945.

17. "Fastest, Farthest, Fighting'est," *Wilmington (NC) Star-News*, December 25, 1945.

18. Stafford, "Flight of the Truculent Turtle," 47.

19. SteelJaw Scribe, "Naval Aviation Centennial: Neptune's Atomic Trident (1950)," *U.S. Naval Institute Blog*, February 6, 2011, https://blog.usni.org/posts/2011/02/06/naval-aviation-centennial-neptunes-atomic-trident-1950.

20. "B-29 Flies Nonstop Honolulu to Manila," *Howard County (NE) Herald*, April 17, 1946.

21. "Queens Sergeant off on Long Pacific Hop," *Long Island Star-Journal*, March 30, 1946.

22. Buck Buchwach, "Fluffy Fuzz V Makes Manila Flight Carrying WAC, Extra Gas Tanks," *Honolulu Advertiser*, March 31, 1946.

23. "High Flyer," 6.

24. "Col. C. S. Irvine Still in News," *St. Paul Phonography*, April 3, 1946.

25. "Col. Irvine Delayed on Coast; Plans Flight to Oahu Tomorrow," *Honolulu Star-Bulletin*, August 24, 1946.

26. "B-29 Sets Fifth Altitude Record," *New York Sun*, May 17, 1946.

27. Huber, "B-29 Box Score," 6.

28. "High Flyer," 6.

29. Thomas D. Davies, as told to Steffan Andrews, "Truculent Turtle Commander Gives His Plane Full Credit," *Dayton (OH) Daily News*, October 4, 1946.

30. "After Four Cadet Years It's Ensign Rankin Yea!," *Sapulpa (OK) Free Press,* June 11, 1937.

31. "PBY-5A Catalina," National Naval Aviation Museum, accessed June 7, 2021, https://www.history.navy.mil/content/history/museums/nnam/explore/collections/aircraft/p/pby-5a-catalina.html.

32. "Black Cats," 114.

33. John V. Young, "Persistence Key Word in Fabled Career," *Sapulpa (OK) Herald,* September 20, 1970.

34. "Navy 'Tells on' Air Hero Who 'Never Said a Word,'" *Albany Knickerbocker News,* August 19, 1944.

35. "Black Cats," 302.

36. "Famous Raiders back in States," *Wilmington (NC) Morning Star,* August 19, 1944.

37. "Ton of Radar on Plane," *Burlington (IA) Hawk-Eye Gazette,* June 7, 1946.

38. "New N.Y.–L.A. Record Set," *Salt Lake Tribune,* May 29, 1946.

39. "Navy Bomber Sets East–West Record," *New York Times,* May 29, 1946. Text corrected from "28 minutes" to the correct time.

40. "Truculent Turtle," *Naval Aviation News,* September 1976.

41. Gulliver, "Truculent Turtle's Excellent Adventure," 45.

42. "Tuning Up at Perth," *New York Sun,* September 19, 1946.

43. George Dixon, Washington Scene, *Tucson Daily Citizen,* October 22, 1946.

5. Superfortress

1. "Lieut. Irvine to Fly Bomber over This City," *St. Paul Phonograph,* April 21, 1937. The newspaper didn't report later whether Irvine had flown the plane low over St. Paul, as he had hoped to do during the flight east to Dayton.

2. Bernard Brookes, "Famous 'Grandad' of the B-29," *Washington Star,* June 25, 1944.

3. "Global Power," advertisement, *Washington Star,* October 13, 1946.

4. Boyne, "B-29's Battle of Kansas," 95.

5. "29 Are Feared Dead in Crash of Bomber on Seattle Plant," *Washington Star,* February 19, 1943.

6. Phillips, "Wichita Builds the B-29," 28.

7. Gurney, *B-29 Story,* 10–11.

8. "Battle of Kansas," 482.

9. Price, "Air Force Gives Birth."

10. Neal, "Add Miles," 43.

11. Price, "Air Force Gives Birth."

12. Boyne, "B-29's Battle of Kansas," 96.

13. "Clarence Irvine Is Elevated to Colonel," *Howard County (NE) Herald,* August 19, 1942.

14. "Col. Irvine, Nelson Man, Awarded Legion of Merit," *Lincoln Nebraska State Journal,* December 26, 1944.

15. USAFB, "Lieutenant General Clarence S. Irvine."

16. Morrison, *Birds from Hell,* 149.

17. Ortensie, "'Battle of Kansas.'"

18. Denis Warner, "'An Hour Ago We Knocked Hell out of Tokio,'" *Sydney Sun*, November 26, 1944.

19. Arvid Shulenberger, "Splash One Dreamboat," *Flight Journal*, October 2001.

20. Boyle, "This Dreamboat Can Fly!," 86.

21. USAFB, "Lieutenant General Clarence S. Irvine."

22. Gladwell, *Bomber Mafia*, 8.

23. Alfonso A. Narvaez, "Gen. Curtis LeMay, an Architect of Strategic Air Power, Dies at 83," *New York Times*, October 2, 1990.

24. Curtis E. LeMay, *Mission with LeMay*, in William Anderson, "A General Tells His Reckoning," *Chicago Tribune*, September 21, 1965. LeMay later said he hadn't written the passage in his memoirs but had let it pass unchecked.

25. Gladwell, *Bomber Mafia*, 135.

26. Coffey, *Iron Eagle*, 154.

27. LeMay and Yenne, *Superfortress*, 114.

28. Tillman, *LeMay*, chap. 5.

29. Elmont Waite, "Nagoya War Plants, Harbor Installations Suffer Heavy Damage," *Geneva (NY) Daily Times*, March 13, 1945.

30. "Col. Clarence Irvine Receives High Military Decorations," *St. Paul Phonograph*, March 20, 1946.

31. "Clarence S. Irvine," Hall of Valor Project, https://valor.militarytimes.com/hero/45294.

32. "Beverly H. Warren," *American Aviation*, September 1966.

33. "These Are the Crewmen," *Honolulu Star-Bulletin*, October 4, 1946.

34. "On Round-World Flight Plans," *Lincoln Nebraska State Journal*, October 30, 1946.

35. USAFB, "Brigadier General Beverly Howard Warren."

36. "Cairo Flight Waits Weather," *Walla Walla Union-Bulletin*, September 6, 1946.

6. Distances

1. "'Dreamboat' Tested for Hawaii to Egypt Flight," *Honolulu Star-Bulletin*, August 22, 1946.

2. Blair Moody, "B-29 Flight, Hawaii to Cairo, Scheduled for Late This Month," *Honolulu Star-Bulletin*, July 16, 1946.

3. Ted P. Wagner, "J— Army Has Plane That Flew 11,000 Miles Non-Stop, a Record," *St. Louis Post-Dispatch*, September 14, 1945.

4. George McCadden, "Japanese Claim Longer Hop Than Hawaii–Egypt," *Honolulu Advertiser*, August 4, 1946.

5. Douglas J. Ingells, introduction to Hays, "Across the Top," 25–27.

6. John Noble Wilford, "Did Byrd Reach Pole? His Diary Hints 'No,'" *New York Times*, May 9, 1996.

7. "Aviation Sees Polar Region Route from Russia to U.S.," *Buffalo Courier-Express*, July 21, 1935.

8. Thomas and Thomas, *Famous First Flights*, chap. 14.

9. "Science Backs Flight," *San Francisco Examiner*, August 3, 1935.

10. Duffy and Kandalov, *Tupolev*, 69.

11. "Russian Aviators Abandon Flight to California," *Buffalo Courier-Express*, August 4, 1935.

12. "Soviet Polar Fliers Expected to Reach Oakland Airport Soon after Daybreak Today," *Oakland Tribune*, June 20, 1937.

13. "Shy Russian Flyers Hailed by 5,000 at Oakland Field," *San Francisco Examiner*, June 22, 1937.

14. "Radio Network Built Secretly, Guided Fliers," *Binghamton Press*, June 21, 1937.

15. Salnikov, "Handshake across the Arctic Ocean."

16. "'Never Once Lost,' Say Fliers," *San Francisco Examiner*, June 21, 1937.

17. "Soviet Fliers, 63 Hours in Air, Tell of Hop," *Binghamton Press*, June 21, 1937.

18. "Russ Flyers Take No Food, No Sleep," *San Francisco Examiner*, June 21, 1937.

19. Alfred P. Reck, "Dry Oxygen Tank Halted Pole Flight," *Oakland Tribune*, June 22, 1937.

20. National Park Service, "Red Bolt."

21. "Soviet Fliers, 63 Hours."

22. "Congratulations Sent Airmen by Roosevelt," *San Francisco Examiner*, June 21, 1937.

23. "Volga Birdmen Praise Motor," *San Francisco Examiner*, June 22, 1937.

24. "Soviet Polar Fliers Due Here at 6 P.M.; World Praise Given," *Oakland Tribune*, June 21, 1937.

25. "North Pole Soon Just Air Route Way Station," *San Francisco Examiner*, June 21, 1937.

26. "Polar Flights Are Held Practical by Wilkins," *Oakland Tribune*, June 21, 1937.

27. Cecilia Rasmussen, "Three Soviet Fliers' 1937 Happy Landing in a Southland Pasture," *Los Angeles Times*, July 22, 2001.

28. "Soviet Fliers End Record Hop; Land near San Jacinto, Cal.," *New York Sun*, July 14, 1937.

29. "Russian Aviators Land in Pasture near San Jacinto after Record Hop," *Washington Star*, July 14, 1937.

30. "3 Russian Aviators Land in California; Easily Top Record," *New York Times*, July 15, 1937.

31. "Fliers Talk about Flight," *New York Times*, July 15, 1937.

32. "Feat of Airmen Hailed in Soviet," *New York Times*, July 16, 1937.

33. "Soviet Warns 'Enemies' Face Attack by Air," *Washington Star*, July 15, 1937.

34. "Three Planes on Record Attempt," *Sydney Sun*, November 6, 1938.

35. "Bombers Pass India on Try for Record," *New York Times*, November 6, 1938.

36. "British Planes Seeking Record Strike Storm," *New York Sun*, November 5, 1938.

37. "7162 Miles!—Bombers Smash Record," *Sydney Sun*, November 7, 1938.

38. O'Connor, "Flight of the Wellesleys."

39. "British Planes End Flight at Darwin," *New York Times*, November 8, 1938.

40. "Casual R.A.F. Record-Breakers," *Sydney Daily Telegraph*, November 8, 1938.

41. "High Praise," *Sydney Daily Telegraph*, November 8, 1938.

7. Honolulu

1. "Dreamboat Is Being Readied in Seattle," *Honolulu Star-Bulletin*, August 5, 1946.

2. "Plane Repairs Delay Cairo Flight," *Honolulu Advertiser*, August 12, 1946.

3. Irvine, "The Briefing," 2.

4. Eaker letter to Irvine, September 11, 1946, AL-7 Col. Clarence "Bill" Irvine Photo Album.

5. Boyle, "This Dreamboat Can Fly!," 90.

6. "Ready for Flight over North Pole," *Salem (OR) Capital Journal*, August 22, 1946.

7. "Dreamboat to Take Off Here Sept. 1 to 10 on Cairo Flight," *Honolulu Advertiser*, August 24, 1946.

8. "Dreamboat May Seek New Nonstop Record," *San Antonio Light*, October 5, 1946.

9. "'Dreamboat' Will Hop for Hawaii," *Oakland Tribune*, August 25, 1946.

10. "Col. Irvine Now Scheduled to Fly Here Tomorrow," *Honolulu Star-Bulletin*, August 26, 1946.

11. "Takeoff of 'Dreamboat' to Hawaii Is Delayed," *Philadelphia Inquirer*, August 28, 1946.

12. "Dreamboat Due Today as Cairo Hop Nears," *Honolulu Star-Bulletin*, August 31, 1946.

13. "Super Fortress Hops to Hawaii to Prepare for Cairo Flight," *Washington Star*, September 1, 1946.

14. Buck Buchwach, "Dreamboat Gets Final Test Today for Cairo Hop," *Honolulu Advertiser*, September 4, 1946.

15. "'Dreamboat' Gets 100-Hour Inspection at Hickam Field," *Honolulu Star-Bulletin*, September 2, 1946.

16. Buchwach, "Dreamboat Gets Final Test."

17. "Arctic Twilight Hours Will Be 'Blind Spot' on Dreamboat Flight," *Honolulu Star-Bulletin*, September 3, 1946.

18. "Here Are Some Dreamboat Facts and Figures; Plane Tested Today," *Honolulu Star-Bulletin*, September 4, 1946.

19. "Weather Outlook Uncertain for Oahu-to-Cairo Flight," *Honolulu Star-Bulletin*, September 2, 1946.

20. "Dreamboat Is Readied for 10,030-Mile Flight," *Honolulu Advertiser*, September 3, 1946.

8. Perth

1. "Americans Back," *Perth West Australian*, September 6, 1946.

2. Sturma, *Fremantle's Submarines*, 1.

3. "Farewell to Patwing Ten," *Perth West Australian*, November 17, 1944.

4. ENHP, letter to family dated September 5, 1946.

5. "Americans Back."

6. "U.S. Plane May Make Perth to Seattle Flight in One Hop," *Brisbane Telegraph*, September 20, 1946.

7. Moore, "Interview on Experiences in World War II."

8. Horner, "Takeoff of the Truculent Turtle."

9. *Lucky Bag*, 273.

10. "Three Wives 'Sweat It Out' Here as Turtle Wings Way Eastward," *Washington Post*, October 1, 1946.

11. Ann Mittman, "The Long Flight of the 'Turtle,'" *Cocoa Florida Today*, December 19, 1986.

12. *VP-81 Album*, 83.

13. RHTP, letter from Chief of Naval Personnel, August 13, 1946.

14. "New American Plane," *Perth West Australian*, September 19, 1946.

15. "Perth–Seattle Flight Soon," *Perth Daily News*, September 19, 1946.

16. "Navy Plane to Fly 9,000-Mile Non-Stop Test from Australia to U.S. on Endurance Check," *New York Times*, September 19, 1946.

17. "Navy Plane Plans Record Hop," *New York Sun*, September 19, 1946.

18. "The Turtle's Triumph," editorial, *Washington Star*, October 2, 1946.

19. "Navy Outlines 9,000-Mile Australia-to-U.S. Flight," *Amarillo Daily News*, September 19, 1946.

20. "The Neptune," *Perth West Australian*, September 20, 1946.

21. "Flyers Acclaimed in Brisbane after Spanning Pacific," *Washington Star*, June 9, 1928.

22. "The Neptune," *Perth West Australian*, September 21, 1946. The newspaper used several identical headlines in reporting on the P2V.

23. "Aries in Perth," *Perth West Australian*, September 21, 1946.

24. "Airport Visitors," *Perth West Australian*, September 21, 1946.

25. "Hurricane May Delay Perth–Seattle Flight," *Perth Sunday Times*, September 22, 1946.

26. "Lockheed P2V Neptune," 40.

27. "Neptune Leaves Next Week," *Perth Daily News*, September 21, 1946.

28. "Neptune and Aries," *Perth West Australian*, September 23, 1946.

29. "Combined Operations," *Perth West Australian*, September 23, 1946.

9. Gremlins

1. Curtiz, *Casablanca*.

2. "Col. Irvine Has Handpicked Crew for Oahu–Cairo Nonstop Flight," *Honolulu Star-Bulletin*, September 2, 1946.

3. "Arctic Twilight Hours" (see chap. 7, n. 17).

4. "Letters: Dreamboat's Runway," *Popular Science*, February 1947.

5. "Dreamboat Will Avoid City in Its Takeoff," *Honolulu Star-Bulletin*, August 28, 1946.

6. "Col. Irvine Has Handpicked Crew."

7. Keyes Beech, "'Dreamboat' Waits Favorable Skies for Polar Flight," *Oakland Tribune*, September 5, 1946. Beech worked for the *Star-Bulletin* but filed the story as a stringer for the *Chicago Daily News* wire service.

8. "Navigator on Dreamboat Becomes Father," *Honolulu Star-Bulletin*, September 5, 1946.

9. "Dreamboat's Test Flight Satisfactory, Col. Irvine Reports," *Honolulu Advertiser*, September 6, 1946.

10. "Woman Metal Worker Has Role in Dreamboat Flight," *Honolulu Star-Bulletin*, September 6, 1946.

11. "'Dreamboat' on Second Test Hop at Hickam Today," *Honolulu Star-Bulletin*, September 6, 1946.

12. Buck Buchwach, "Dreamboat Hop No Stunt, Gen. Whitehead Avers," *Honolulu Advertiser*, September 7, 1946.

13. Buck Buchwach, "Dreamboat's Final Test Flight Thrills Newsmen," *Honolulu Advertiser*, September 8, 1946.

14. Keyes Beech, "Col. Irvine, Crew and Newsmen Put Okay on Pacusan Dreamboat," *Honolulu Star-Bulletin*, September 9, 1946.

15. Buchwach, "Dreamboat's Final Test."

16. "St. Paul Anxiously Follows" (see chap. 3, n. 8).

17. "Dreamboat in Perfect Form," *Honolulu Advertiser*, September 10, 1946.

18. W. H. Shippen Jr., "World Weather News Vital to Victory Stems from Air Forces Center Here," *Washington Star*, February 11, 1945.

19. Yates, "We Shall Have Weather," 10.

20. Fred Dubois, "Army's 'Operation Stork' Maps Weather from Bay Area to Alaska," *Oakland Tribune*, September 5, 1946.

21. John I. Kent, "Weather Bureau Planes Now Intercept, Check Up on Storms Headed U.S.-ward," *Cohoes (NY) American*, March 3, 1947.

22. "History, 53d Reconnaissance Squadron," 23–24.

23. "Weather Service Will Keep 'Dreamboat' Posted All the Way," *Honolulu Advertiser*, September 5, 1946.

24. "Won't Cross Soviet Territory," *New York Sun*, October 4, 1946.

25. "Dreamboat's Route Changed Due to Weather; Takeoff Soon," *Honolulu Advertiser*, September 24, 1946.

26. Crowson, "Meteorological Aspects," 221.

27. Hubbard, "Arctic Isn't So Tough."

28. "'Dreamboat' Said Ready to Go This Week," *Honolulu Advertiser*, September 15, 1946.

29. "Phila. Youth in Navy Mentioned for Valor," *Philadelphia Inquirer*, November 19, 1918.

30. "Dreamboat Beats Storm, by Flying near Yukon on Top-of-World Hop," *Philadelphia Inquirer*, October 5, 1946.

31. "B-29 Crew Bleary-Eyed after Flight from Guam," *Philadelphia Inquirer*, November 24, 1945.

32. "Irvine's Pending Flight to Cairo," *Howard County (NE) Herald*, September 4, 1946.

33. "Army and Navy Merge Services for 'Dreamboat,'" *Honolulu Star-Bulletin*, September 14, 1946.

34. "PACUSAN Dreamboat," *Stars and Stripes*, Pacific, October 6, 1946.

35. "Gas Leak Delays Departure of Dreamboat at Least 5 Days," *Honolulu Advertiser*, September 12, 1946.

36. "Dreamboat Crew, Delayed Again, Deny They Plan to Be Kamaainas," *Honolulu Star-Bulletin*, September 12, 1946.

37. "Dreamboat Takeoff Expected Next Week after Repairs," *Honolulu Advertiser*, September 14, 1946.

38. "Oahu–Cairo Hop Delayed by Gas Leak," *Honolulu Star-Bulletin*, September 17, 1946.

39. Keyes Beech, "This Little World," *Honolulu Star-Bulletin*, October 9, 1946.

40. "Dreamboat Flight off Indefinitely; New Gas Tank Leak," *Honolulu Advertiser*, September 18, 1946.

41. U.S. War Department, "'Pacusan Dreamboat' Pilot Wishes Luck to Naval Fliers," September 30, 1946, *Press Releases*.

42. "A Healthy Rivalry," *Honolulu Advertiser*, September 22, 1946.

43. Howard D. Case, "Down to Cases," *Honolulu Advertiser*, September 29, 1946.

44. "Dreamboaters' Contribution," editorial, *Honolulu Advertiser*, October 8, 1946.

10. Navigators

1. "Bausch & Lomb News Release," 214.

2. "Navy Flight Tests Pressure Navigation," *Aviation News*, September 30, 1946.

3. "Tiff City Navigator Is Awarded the D.F.C.," *Joplin (MO) Globe*, September 21, 1945.

4. Hays, "Wanted: 1000 Navigators," 41.

5. Grace Wing, "On the Wing," *Miami Daily News*, November 8, 1946.

6. "Dream Comes True as Maj Brothers Sees Son," *Knoxville (TN) Journal*, November 16, 1946.

7. "Knox Flying Maj. James Brothers Makes Paris-Chicago Trip Home with Gen. Clark," *Knoxville (TN) News-Sentinel*, June 1, 1945.

8. Hays, "Across the Top of the World," 92.

9. "Shifting North Pole a Mystery to Scientists," *Townsville (Queensland) Daily Bulletin*, December 30, 1946.

10. "Soviet Aviators Pass Halfway Mark on Flight to U.S. via Pole," *Washington Star*, June 19, 1937.

11. Shannon Hall, "The North Magnetic Pole's Mysterious Journey across the Arctic," *New York Times*, February 4, 2019.

12. Leman, "Magnetic North Pole." The article relays in layman's terms research by Livermore, Finlay, and Bayliff in "Recent North Magnetic Pole."

13. Davies, "Thank the Planet's."

14. Buchwach, "Dreamboat Gets Final Test" (see chap. 7, n. 14).

15. Max B. Cook, "New Type of Compass Unerringly Guides United Nations Fliers," *El Paso Herald-Post*, October 13, 1943.

16. "New Gyro Compass 'Navigates' Planes," *New York Times*, February 12, 1944.

17. "Test Tube Genii Help Win War," *Spokane Spokesman-Review*, January 3, 1944.

18. Lewis, "Navigational Stars," 211.

19. George McCadden, "'Dreamboat' to Use New Streamlined Navigation," *Honolulu Advertiser*, September 5, 1946.

20. Halliday, "Aries Flights of 1945."

21. Grierson, *Challenge to the Poles*.

22. "Flights over North Pole," *London Guardian*, May 15, 1945.

23. Broughton, "Aries Flights," in *History of Navigation*, 75.

24. "Oxygen Needs Rise in Arctic Flying," *Polar Times*, December 1946.

25. Maclure, "Technical Aspects," 109.

26. "Science: The Aries," *Time*, June 4, 1945.

27. Broughton, "Aries Flights," 77.

28. "Important Discoveries of Aries' Flight," *London Guardian*, May 28, 1945.

29. E. W. Anderson, "What Is It Like at the North Pole?," *Vancouver Province*, September 1, 1945.

30. McKinley, "Arctic Flights of *Aries*," 98.

31. "Obituary: Air Vice-Marshal David McKinley," *London Telegraph*, May 2, 2002.

11. JATO

1. "The Neptune," *Perth West Australian*, September 24, 1946.

2. Boyne, "Von Karman's Way."

3. Daso, *Architects of American Air Supremacy*, 70.

4. Frederick Johnsen, "JATO Pushed Performance," *General Aviation News*, February 7, 2019, https://generalaviationnews.com/2019/02/07/jato-pushed-performance/.

5. Taylor, "Jatos Get 'Em Up."

6. "Taberling [*sic*] Gives Personal Story of Trans-Pacific Hop," *Wardial*, March 21, 1947.

7. "Naval Air Forces Will Get Rotation to North Pacific," *Honolulu Advertiser*, September 1, 1946.

8. "Jet Take-off," *Perth West Australian*, September 25, 1946.

9. "Rockets for Take-off on Record Flight," *Perth Daily News*, September 24, 1946.

10. Hernan, "Truculent Turtle."

11. Crowe and Chanoff, *Line of Fire*, 244.

12. Polmar, "Historic Aircraft," 12.

13. Crowe and Chanoff, *Line of Fire*, 244.

14. "Best Wishes," *Perth West Australian*, September 25, 1946.

15. "The Neptune," *Perth West Australian*, September 26, 1946.

16. "Neptune's Start May Be Delayed," *Perth Daily News*, September 25, 1946.

17. "The Neptune," *Perth West Australian*, September 27, 1946.

18. "Neptune Goes to Pearce," *Perth Daily News*, September 26, 1946.

19. "Newsboy Is Crew's Friend," *Perth Daily News*, September 27, 1946.

20. "Moral Pointed," *Perth West Australian*, October 5, 1946.

21. News and Notes, *Perth West Australian*, December 26, 1946.

22. "Turtle's Technician Was 'Disappointed' by Epic Hop," *Honolulu Star-Bulletin*, October 8, 1946.

23. "Made Airman Hop," *St. George Balonne Beacon*, November 7, 1946.

24. "Navy Plane Sets Mark of 11,236 Miles," *Los Angeles Times*, October 2, 1946.

25. "Neptune Off," *Perth West Australian*, September 30, 1946.

26. "Neptune's Start Delayed," *Perth Daily News*, September 27, 1946.

27. "Neptune Waits on Weather," *Sydney Sun*, September 27, 1946.

28. *Aerology Operational Analysis*, 1.

29. Harding, "U.S. Naval Weather Service," in *6th U.S. Navy Symposium*, 17.

30. "Rehoboth's Part," *Perth West Australian*, September 26, 1946.

31. *Aerology Operational Analysis*, 2.

32. "Flight Postponed," *Bathurst National Advocate*, September 28, 1946.

33. "Neptune's Take-off Today Is Probable," *Perth Sunday Times*, September 29, 1946.

34. "Neptune Still Delayed," *Perth Daily News*, September 28, 1946.

35. "Dreamboat Takeoff Time Moved to Early Morning," *Honolulu Star-Bulletin*, September 13, 1946.

36. "Dreamboat Set for Long Hop; Up to Weather," *Honolulu Advertiser*, September 25, 1946.

37. "Pacusan Takeoff Is Postponed at Last Minute," *Honolulu Star-Bulletin*, September 28, 1946.

12. The Pacific

1. "Turtle Ends Record Flight," *New York Sun*, October 1, 1946.

2. Patrol Squadron Two Association, "Tales of the 'Truculent' Turtle."

3. *Aerology Operational Analysis*, 2.

4. Davies and Andrews, "Truculent Turtle Commander" (see chap. 4, n. 29).

5. Reid, "I Rode the Turtle," 272.

6. Davies and Hanson, "In 'Operation Turtle,'" 71.

7. "Plane Takes Off on Long Trip to U.S.," *Sydney Daily Telegraph*, September 30, 1946.

8. Foster, "Science on Wings."

9. "Neptune Leaves Next Week" (see chap. 8, n. 27).

10. "News from Australia," *The Aeroplane*, January 3, 1947.

11. "Navy Plane off from Perth on Australia-to-U.S. Flight," *New York Times*, September 30, 1946.

12. "Dog and Neptune Kangaroo Clash," *Perth Daily News*, September 30, 1946.

13. George A. Scott, "Warning, 'Never Forget,' Is Printed on Austrian Stamps," *Columbus (OH) Dispatch*, October 27, 1946.

14. "Goodwill Message," *Perth West Australian*, October 1, 1946.

15. "Distance Plane on Way to U.S.," *Philadelphia Inquirer*, September 30, 1946.

16. "Non-stop Plane 'Turtle' Passes Milne Bay," *Perth Daily News*, September 30, 1946.

17. Rankin, "It Was: One Long Hop," 23.

18. Rankin, "One Long Hop," 23.

19. Gulliver, "Truculent Turtle's Excellent Adventure," 46.

20. Rankin, "Narrowing Horizons," 336.

21. Mittman, "Long Flight" (see chap. 8, n. 11).

22. "Neptune Off" (see chap. 11, n. 25).

23. "Jitters and Jets," in *Of Men and Stars*, 3.

24. R. H. Tabeling, "Trip Was Perfect, Says Crew Member of 11,000-Mile Flight," *Columbus (OH) Dispatch*, October 1, 1946.

25. "Jato Assures Turtle Record," *Palacios (TX) Beacon*, November 28, 1946.

26. *Aerology Operational Analysis*, 2.

27. "Washington Prepares for Plane," *New York Times*, September 30, 1946.

28. Tabeling, "Trip Was Perfect."

29. "Truculent Turtle's Crew Notes Epic Navy Flight while Celebrating Silver Anniversary," *Van Nuys Valley News*, October 14, 1971.

30. "Battle of the Coral Sea," Royal Australian Navy, accessed October 14, 2020, https://www.navy.gov.au/history/feature-histories/battle-coral-sea.

31. "Navy Flyers Pass Halfway Point in Hop to U.S.," *New York Sun*, September 30, 1946.

32. *Aerology Operational Analysis*, 2.

33. "Navy's Flying 'Turtle' over Coral Sea on Record-Seeking Nonstop Hop," *Honolulu Advertiser*, September 30, 1946.

34. "Turtle Ends Record Flight."

35. "Grilled Steaks on Record Flight," *Sydney Sun*, October 3, 1946.

36. W. H. Shippen Jr., "AAF Effort to Topple Mark of Truculent Turtle Awaited," *Washington Star*, October 2, 1946.

37. Mervin Roland, "Columbus' Value as World Air Terminal Is Seen by 'Turtle' Crew Landing Here," *Columbus (OH) Citizen*, October 2, 1946.

38. *Aerology Operational Analysis*, 2.

39. Rankin, "It Was: One Long Hop," 23.

40. "Air Triumph," *Perth West Australian*, October 3, 1946.

41. Reid, "I Rode the Turtle," 272.

42. "Neptune's Flight," *Perth West Australian*, October 1, 1946.

43. "'Non-Stop' Plane Has Flown More Than 5300 Miles," *Newcastle Sun*, October 1, 1946.

44. "Neptune Is Bound for Bermuda," *Perth Daily News*, October 1, 1946.

45. William Kroger, "Lockheed Neptune Record Flight Reveals Efficiency of New Design," *Aviation News*, October 7, 1946.

46. Stafford, "Flight of the Truculent Turtle," 46.

47. Tabeling, "Trip Was Perfect."

48. *Aerology Operational Analysis*, 2. All references to a navy report refer to this document.

13. Landfall

1. Shippen, "AAF Effort to Topple Mark" (see chap. 12, n. 36).

2. "Navy Plane May Try for Bermuda," *New York PM*, October 1, 1946.

3. "Navy 'Turtle' Breaks Record," *San Bernardino Sun*, October 1, 1946.

4. "Distance Plane Reaches U.S., Breaks Record," *Philadelphia Inquirer*, October 1, 1946.

5. "'Turtle' Sets World Mark," *San Francisco Examiner*, October 1, 1946.

6. "Truculent Turtle Lands at Columbus, Setting 11,237-Mile Record," *Washington Star*, October 1, 1946.

7. "Neptune's Flight," *Perth West Australian*, October 2, 1946.

8. "Navy Plane Sets World Mark," *Santa Rosa Press Democrat*, October 1, 1946.

9. Stafford, "Flight of the Truculent Turtle," 47.

10. "Jitters and Jets," in *Of Men and Stars*, 4.

11. *Aerology Operational Analysis*, 3.

12. "Truculent Turtle's Crew Notes" (see chap. 12, n. 29).

13. "Turtle Sets Mark, Lands at Columbus," *Circleville (OH) Herald*, October 1, 1946.

14. "'Turtle' over U.S., Heads for Bermuda," *Los Angeles Times*, October 1, 1946.

15. "'Turtle' over U.S."

16. William Shakespeare, *The Tempest*, act 1, scene 2.

17. "Sailors Feel Safe."

18. Drew Pearson, Washington Merry-Go-Round, *Rome (NY) Sentinel*, October 9, 1946.

19. "Record Seeking Plane Roars Past Midway on Way to Capital," *Washington Star*, September 30, 1946.

20. Rockwell, "So You Want to See."

21. "Three Wives 'Sweat It Out'" (see chap. 8, n. 10).

22. "Cays Is Home for 'Truculent Turtle' Flight Record Setter of '46," *Coronado Journal*, August 29, 1974.

23. "Parents Here Await Word of Flight Result," *Sapulpa (OK) Herald*, September 30, 1946.

24. Roland, "Columbus' Value" (see chap. 12, n. 37).

25. "'Truculent Turtle' Smashes," 18.

26. "'Turtle' Sets Flight Record," *Rome (NY) Sentinel*, October 1, 1946.

27. Mittman, "Long Flight" (see chap. 8, n. 11).

28. "'Truculent Turtle' Wings over State," *Casper Tribune-Herald*, October 1, 1946.

29. "Await Plane Here in Vain," *Des Moines Tribune*, October 1, 1946.

30. "Wings across Nebraska," *Lincoln (NE) Star*, October 1, 1946.

31. "New Records," editorial, *Rock Island (IL) Argus*, October 2, 1946.

32. Mary Little, Airglances, *Des Moines Tribune*, October 2, 1946.

33. "Await Plane Here."

34. Davies and Andrews, "Truculent Turtle Commander" (see chap. 4, n. 29).

35. Browsing Around, *Ottumwa Courier*, October 1, 1946.

36. "Navy Plane Sets Mark of 11,236 Miles" (see chap. 11, n. 24).

37. "Record Breaker over Lafayette," *Lafayette (IN) Journal and Courier*, October 1, 1946.

38. "Navy Plane from Australia Lands at Columbus," *Defiance Crescent-News*, October 2, 1946.

39. Frederick Graham, "11,236-mile Record Set as Navy Plane Lands in Columbus," *New York Times*, October 2, 1946.

40. "From Way Down Under," editorial, *Akron Beacon Journal*, October 2, 1946.

41. Gerald Tebben, "Columbus Mileposts: Oct. 1, 1946—Plane Carries Kangaroo into Town," *Columbus (OH) Dispatch*, October 1, 2012.

42. "Cargo of Prize Cows Flown from California," *Dayton Daily News*, October 2, 1946.

43. "Truculent Turtle Lands in Ohio to End Longest Hop," *Philadelphia Inquirer*, October 2, 1946.

44. Weekly News Analysis, *Shafter (CA) Press*, October 17, 1946.

45. "Bermuda 'a Cinch,'" *Perth Daily News*, October 2, 1946.

46. "Turtle's Technician" (see chap. 11, n. 22).

47. Buck Buchwach, "Russ Refuse US Permit for Flight of Turtle," *Honolulu Advertiser*, October 9, 1946.

48. "Turtle Wins Long Race," editorial, *New York Times*, October 2, 1946.

49. Barbara Carmen, "51 Years Ago, World Turned Its Attention to Plane, Columbus," *Columbus (OH) Dispatch*, October 3, 1997.

50. Richard Cull, "'Turtle' Crewmen Regard Hop as 'Just Long Patrol Flight,'" *Dayton Daily News*, October 2, 1946.

51. "Record Non-stop Navy Flight Ends Here," *Columbus (OH) Dispatch*, October 1, 1946.

52. "Turtle Battled Ice, Headwinds to Shatter Air Record," *Sydney Sun*, October 2, 1946.

53. Tebben, "Columbus Mileposts."

54. Graham, "11,236-mile Record Set."

55. Tabeling, "Trip Was Perfect" (see chap. 12, n. 24).

56. "Plane Flies 11,237 Miles Non-Stop from Australia to Ohio to Set Record," *Joplin (MO) Globe*, October 2, 1946.

57. Roland, "Columbus' Value."

58. "Record Non-stop Navy Flight Ends Here."

59. "Kangaroo Is Suggested as Dinner Steak," *Jefferson City (MO) Capital News*, October 2, 1946.

60. "Air Triumph" (see chap. 12, n. 40).

61. Davies and Andrews, "Truculent Turtle Commander."

62. Fletcher Knebel, "Navy Crew of Turtle Decorated," *Cleveland Plain Dealer*, October 2, 1946.

63. "Rankin Phones Parents Regarding Plane Flight," *Sapulpa (OK) Herald*, October 2, 1946.

64. "Greeted by Wives," *Joplin (MO) Globe*, October 2, 1946.

65. "Crew Cocky and Alert after Record Flight," *Dayton Herald*, October 2, 1946.

66. RHTP, untitled manuscript, "Newsbureau [*sic*] Lockheed Burbank," 5.

67. "Flight of the Turtle," *Ohio State Journal*, editorial, October 3, 1946.

68. Wylie, *Radio and Television Writing*, 200.

69. News and Notes, *Perth West Australian*, December 26, 1946.

70. "Joey: World Citizen," *Nashua Telegraph*, quoting *Christian Science Monitor*, October 5, 1946.

71. Ad Schuster, Other Fellow, *Oakland Tribune*, October 3, 1946.

14. The Pole

1. "Irvine Praises Turtle's Airmen for Record Hop," *Honolulu Advertiser*, October 2, 1946.

2. U.S. War Department, "Pacusan Dreamboat and Crew Returning to Washington," October 16, 1946, *Press Releases*.

3. "Twin Engine Navy Plane Sets New Record of 11,822 Miles," *Honolulu Star-Bulletin*, October 1, 1946.

4. "Dreamboat Will Carry On with Cairo Flight Plans," *Honolulu Star-Bulletin*, October 2, 1946.

5. "'Dreamboat's' Crewmen Home," *Dayton Herald*, October 19, 1946.

6. "Dreamboat Tested," *Honolulu Advertiser*, October 3, 1946.

7. "Now It's the Dreamboat's Turn," *Honolulu Star-Bulletin*, editorial, October 3, 1946.

8. Crowson, "Meteorological Aspects," 224.

9. Jo Driskell, "Former Shoe Clerk Returns from History-Making Flight," *Moberly (MO) Monitor-Index*, November 15, 1946.

10. Moray Epstein, "History Books' Pages Rustle in Wash of the Dreamboat's Props," *Honolulu Star-Bulletin*, October 4, 1946.

11. "Dreamboat Set for Hop Today," *Honolulu Advertiser*, October 4, 1946.

12. Don Whitehead, "'Dreamboat' Starts Top of World Hop," *San Pedro (CA) News-Pilot*, October 4, 1946.

13. "Advertiser's First Egypt Edition on Dreamboat," *Honolulu Advertiser*, October 5, 1946.

14. William D. Eberhart, "Army Fort Off on Pole Flight to Cairo, Egypt," *Cohoes (NY) American*, October 4, 1946.

15. Keyes Beech, "Dreamboat Now Close to Sitka," *Honolulu Star-Bulletin*, October 4, 1946.

16. "Thumbnail Sketches of Men Who Are Riding the Winds to Egypt," *Honolulu Star-Bulletin*, October 5, 1946.

17. "Inside the Dreamboat," 92.

18. Beech, "Dreamboat Now Close."

19. "St. Paul's Irvine Promises [to] Give Up Distance Flights," *Lincoln (NE) Star*, October 7, 1946.

20. "Co-Pilot of the Dreamboat's Record Flight Is Son-In-Law of Omaha Druggist," *Midwestern Druggist* 22, no. 1 (November 1946): 23, 30.

21. "Dreamboat's Commander Has High Praise for Sgt. Fish," *Berkshire County (MA) Eagle*, October 25, 1946.

22. "Husband of Pittsfield Girl Crew Chief on Dreamboat," *Berkshire County (MA) Eagle*, October 5, 1946.

23. "A Sergeant from Pittsfield Rides Dreamboat over Arctic," *Albany Knickerbocker News*, October 5, 1946.

24. "A.C.S. Reports Plane's Progress," *Sitka Sentinel*, October 4, 1946.

25. "The 'Dreamboat' Flight," 66.

26. "Dreamboat Beats Storm" (see chap. 9, n. 30).

27. "M. Sgt. Gray Vasse Praised for Work on Pacusan Flight," *Moberly (MO) Monitor-Index*, October 23, 1946.

28. "Huntsville Airman in Crew of B-29 Seeking Flight Record," *Moberly (MO) Monitor-Index*, August 8, 1946.

29. "Safer to Look for More Zeros," *Moberly (MO) Monitor-Index*, January 6, 1943.

30. "Dreamboat's Commander Has High Praise."

31. "Sgt. Fish of 'Dreamboat' Crew Is Decorated at Washington," *Appleton Post-Crescent*, October 18, 1946.

32. "Air Force Sergeant Finishes 5 Years Testing B-36s in Flight," *Appleton Post-Crescent*, February 20, 1952.

33. Huber, "Over the Top," 5.

34. George L. Peterson, "Don't Be Optimistic about Ride to Alaska," *Minneapolis Star*, October 15, 1946.

35. "Dreamboat Passes Pole," *Honolulu Advertiser*, October 5, 1946.

36. "Thumbnail Sketches."

37. Maurice Aboaf, "North Pole Not Where Shown on Map," *Egyptian Gazette*, October 7, 1946.

38. Harris Peel, "'Dreamboat' Crew Wants Another Try at Distance Mark," *Stars and Stripes*, European edition, October 12, 1946.

39. "Yesterday's Dreamers," editorial, *Christian Science Monitor*, reprinted in *Herkimer (NY) Telegram*, October 9, 1946.

40. Max Boyd, "Flight Establishes Arctic Air Path as Feasible for Plane Operations," *Syracuse Post-Standard*, October 7, 1946.

41. "Plane Passes North Pole on 10,300-Mi. Hop," *Brooklyn Eagle*, October 5, 1946.

42. Huber, "Over the Top," 5.

43. Hays, "Across the Top," 96.

44. Two articles published under Hays's name offer a slightly different position for the magnetic North Pole the day of the *Dreamboat*'s flight. In "Across the Top" in *Air Trails*, it is 73 degrees north, 93 degrees west. In "Report on Navigation" in *Navigation*, it is 73 degrees north, 95 degrees west. The latter is more probably correct.

45. Waldemar Kaempffert, Science in Review, *New York Times*, October 13, 1946.

46. "Seneca Man on Dreamboat Flight," *Joplin (MO) Globe*, November 2, 1946.

47. Driskell, "Former Shoe Clerk."

15. Downhill

1. Hays, "Across the Top," 96.

2. Anderson, "Letters."

3. "Plane Passes North Pole" (see chap. 14, n. 41).

4. "Hawaii B-29 Reaches Cairo after Hop over Top of World," *Philadelphia Inquirer*, October 6, 1946.

5. "Dream Comes True" (see chap. 10, n. 6).

6. Although its crews flew from various bases, the Fifty-Third was headquartered at Grenier Field in Manchester. A local newspaper had reported that the squadron "will be 'flying the weather' ahead of the Dreamboat from the eastern Canadian Arctic to beyond Iceland." Reg Abbott, "B-17s to Fly Region ahead of B-29," *Manchester Leader*, September 5, 1946.

7. Reg H. Abbott, "A Ride with Daring Men Who Pave Daily Air Roads over Atlantic Ocean," *Elmira Star-Gazette*, April 10, 1947. Colonel Irvine transposed the squadron's number and mistakenly recalled it as the Thirty-Fifth in *Popular Science*.

8. "Dreamboat Lands in Cairo," *Manchester Guardian*, October 7, 1946.

9. "Dreamboat Flies over Crete on Last Lap of Trip," *Washington Star*, October 6, 1946.

10. "'Dreamboat' Finds Fighters Napping," *Melbourne Age*, October 8, 1946.

11. "RAF Planes Miss Dreamboat but 'Ham' Picks It Up," *Brooklyn Eagle*, October 7, 1946.

12. "Pacusan Dreamboat Gets to Cairo Goal," *Deadwood Pioneer-Times*, October 6, 1946.

13. "Dreamboat Arrives in Cairo!," *Honolulu Advertiser*, October 6, 1946.

14. "'Dreamboat's' Crewmen Home" (see chap. 14, n. 5).

15. Boyd, "Flight Establishes" (see chap. 14, n. 40).

16. "'Dreamboat' Nearing End of Journey," *Lincoln Sunday Journal and Star*, October 6, 1946.

17. U.S. War Department, "AAF-Developed North African Air Terminals Being Deactivated," March 14, 1947, *Press Releases*.

18. "Dreamboat Arrives in Cairo!"

19. "Dreamboat's Feat Hailed as Proving U.S. Air Defense," *Lincoln Nebraska State Journal*, October 7, 1946.

20. "Dreamboat Ends Global Hop with Gas Nearly Gone," *Philadelphia Inquirer*, October 7, 1946.

21. "Dreamboat Lands at Cairo after 39 Hours," *Sitka Sentinel*, October 7, 1946.

22. "Dreamboat Crew Finds Magnetic Pole 200 Miles off Location," *Wilmington Star*, October 7, 1946.

23. "Dreamboat's Feat Hailed."

24. Driskell, "Former Shoe Clerk" (see chap. 14, n. 9).

25. "Knox Navigator of Dreamboat Hop Plans to Be Pilot," *Knoxville News-Sentinel*, November 15, 1946.

26. "St. Paul's Irvine Promises" (see chap. 14, n. 19).

27. "'Dreamboat' Nearing End."

28. "'Thank Goodness!' Says Mrs. Warren," *Lincoln Nebraska State Journal*, October 7, 1946.

29. Walter Collins, "Pacusan Hop May Open New Route across Roof of World," *Honolulu Advertiser*, October 7, 1946.

30. "King Farouk Sends Thanks," *Honolulu Advertiser*, October 11, 1946.

31. "Servicemen, Civilians at Tinker Celebrate Anniversary of Air Force," *Oklahoma City Daily Oklahoman*, September 19, 1962.

32. "Distance Measurement Corrected," *New York Times*, October 8, 1946.

33. Hamlin, Report from Washington.

34. "Pacusan B-29 Crew Claims Speed Record," *Lincoln Star*, October 8, 1946.

35. Gill Robb Wilson, "Crews Produce Perfect Planes," *Oklahoma City Daily Oklahoman*, October 28, 1946.

16. Blue Skies

1. Drew Pearson, Washington Merry-Go-Round, *San Bernardino (CA) Sun*, October 11, 1946. Ruth Reid is missing from the Rose Garden photographs, but her name appears on President Truman's White House itinerary.

2. "Truculent Turtle," *Perth West Australian*, November 7, 1946.

3. "Famed Flyer to Visit Sapulpa," *Sapulpa (OK) Herald*, October 11, 1946.

4. Drew Pearson, Washington Merry-Go-Round, *Buffalo Courier-Express*, October 12, 1946.

5. "Southland Paid Visit by Truculent Turtle," *Los Angeles Times*, October 15, 1946.

6. "National Air Clinic Spurns Rankin Plea," *Oklahoma City Daily Oklahoman*, October 18, 1946.

7. Rankin, "Narrowing Horizons," 335.

8. Peel, "'Dreamboat' Crew Wants" (see chap. 14, n. 38).

9. "Dreamboat Linked to South Pole Hop," *Philadelphia Inquirer*, October 14, 1946. Karachi was not yet part of an independent Pakistan.

10. "On Round-World Flight Plans" (see chap. 5, n. 34).

11. "Pacusan Flight Not Claimed as a Record," *Lincoln Nebraska State Journal*, October 17, 1946.

12. "Sgt. Fish of 'Dreamboat' Crew" (see chap. 14, n. 31).

13. "The Record Breaking Pacusan Dreamboat Arrives in the Capital," *New York Times*, October 18, 1946.

14. "Dreamboat Lands in Washington," *Philadelphia Inquirer*, October 18, 1946.

15. Ken Clark, "Port Your Helm! Port Your Helm!," *New York PM*, October 18, 1946.

16. "The President Charting the Course of the Pacusan Dreamboat Yesterday," *New York Times*, October 19, 1946.

17. "New York Welcomes Crew of the Pacusan Dreamboat," *New York Times*, October 26, 1946.

18. "100,00 Here Hail Dreamboat Crew," *New York Sun*, October 25, 1946.

17. Commercial Air

1. Stefansson, "Arctic as an Air Route," 205.

2. "Turtle's Flight Leads Navy to Order Planes," *Binghamton Press*, October 2, 1946.

3. "Gen Spaatz Calls Polar Hop Epochal," *New York Times*, October 7, 1946.

4. "Across the Arctic Ice Cap," editorial, *Philadelphia Inquirer*, October 8, 1946.

5. "From Australia to Ohio," editorial, *Milwaukee Journal*, October 2, 1946.

6. "Obit for Oceans," editorial, *New York Post*, October 3, 1946.

7. "Dreamboat over the World," editorial, *New York Times*, October 7, 1946.

8. "Dreamboat's 10,925-Mile Hop Proves a Military Theory," *New York PM*, October 7, 1946.

9. "Pacusan Dreamboat Makes Cairo," editorial, *Rome (NY) Sentinel*, October 7, 1946.

10. "Shows Range of Planes," editorial, *Elmira Star-Gazette*, October 3, 1946.

11. "'Revolution in the Airways?,'" *Key West Citizen*, May 30, 1947.

12. Russell Owen, "The Artic: The North Polar Region Is Likely to Become a Great Commercial Air Route," *New York Times*, January 12, 1947.

13. Clarence S. Irvine, "Flying over Pole Cuts Inter-Continental Time Half, and Will Be Safe as in U.S., Says Dreamboat Skipper," *Albuquerque Journal*, October 18, 1946. Like Roy Tabeling's first-person account of the *Truculent Turtle*'s flight ("Trip Was Perfect," chap. 12, n. 24), Irvine's piece about the *Dreamboat* ran under an International News Service byline. The International News Service likely composed both pieces.

14. "It's a Small World."

15. "Wild Blue Yonder," editorial, *Lincoln Sunday Journal and Star*, October 6, 1946.

16. "Dreamboat Pilots Says Flight Blazed Trail for London-to-Japan Travel in 28 Hours," *Elmira Star-Gazette*, October 7, 1946.

17. "Test Flight: Over Arctic," *Geneseo (NY) Livingston Republican*, October 17, 1946.

18. "Transpolar Air Route Possible, Dreamboat Radio Officer Says," *Washington Star*, October 7, 1946.

19. "'Dream's' Pilot Predicts Easy Polar Flights," *Washington Post*, October 18, 1946.

20. Frederick Graham, "Aviation: Airline Spokesmen See Few Commercial Benefits from Record Military Flight," *New York Times*, October 6, 1946.

21. "America by Air."

22. Bliss K. Thorne, "Aviation: Europe via the Arctic," *New York Times*, November 14, 1954.

23. "Up and over the Top of the World with S-A-S," advertisement, *Los Angeles Times*, November 14, 1954.

24. "Take the 'Short Cut' to Europe," advertisement, *San Francisco Examiner*, August 12, 1957.

25. Drescher, "Why Airplanes Sometimes Fly."

26. Craig R. Whitney, "Soviet Frees Last 2 in Korean Plane Case," *New York Times*, April 30, 1978.

27. Robert D. McFadden, "U.S. Says Soviet Downed Korean Airliner; 269 Lost; Reagan Denounces 'Wanton' Act," *New York Times*, September 2, 1983.

28. Scott Shane, "Roundabout Route Is Shortest," *Baltimore Sun*, December 23, 1999.

29. Anton Troianovski, "Russia Rejects Some Flight Plans, as Belarus Grows More Isolated," *New York Times*, May 27, 2021.

30. Matthew L. Wald, "Study Finds Air Route over North Pole Feasible for Flights to Asia," *New York Times*, October 11, 2000.

31. Mike Dunham, "Russians Search for Lost '30s-Era Soviet Bomber in Alaska," *Anchorage Daily News*, September 28, 2016.

32. Michael Finneran, "Thousand-fold Rise in Polar Flights Hikes Radiation Risk," NASA, February 18, 2011, https://www.nasa.gov/centers/langley/science/polar-radiation.html.

33. Brad Plumer, "Want to Slow Arctic Melting? Stop Flying over the North Pole," *Washington Post*, December 4, 2012.

34. Jacobson et al., "Effects of Rerouting Aircraft," 723.

35. Jeremy Hsu, "Do Airplane Contrails Add to Climate Change? Yes, and the Problem Is about to Get Worse," *NBC News*, July 28, 2019, https://www.nbcnews.com/mach/science/do-airplane-contrails-add-climate-change-yes-problem-about-get-ncna1034521.

36. Cordes, "Over the Top."

Epilogue

1. "Jet Pilots Assault World Speed Records as Aircraft Show Opens," *Mansfield (OH) News-Journal*, November 15, 1946.

2. "'Truculent Turtle' to Carry Byrd to Antarctica," *Binghamton Press*, December 14, 1946.

3. "'Truculent Turtle' May Join Outfit," *New York Times*, January 15, 1947.

4. Bart Barnes, "Adm. Thomas D. Davies, Navigation Expert, Dies," *Washington Post*, January 26, 1991.

5. W. P. Saphir, *New Castle News*, October 11, 1946. Irvine's date of birth is unclear. His veteran's grave marker and many sources and biographies list it as December 16, 1898. Census records and his World War I draft registration card, however, indicate that it was one year earlier.

6. "General Irvine Dies," *St. Paul Phonograph*, September 17, 1975.

7. "Crossfire in Battle with Convicts Perils Biggs Man," *El Paso Herald-Post*, January 5, 1950.

8. Leonard Sloane, "James Kerr, 77, Chief Executive Who Led Avco as Conglomerate," *New York Times*, February 13, 1995.

9. Burt A. Folkart, "James Kerr; Returned Ailing Avco Corp. to Prosperity," *Los Angeles Times*, February 18, 1995.

10. Mark Sarchet, "Tinker's Last B-29 Leaving," *Oklahoma City Daily Oklahoman*, July 8, 1956.

11. Walker S. Buel and Fletcher Knebel, "Ohio under the Dome," *Cleveland Plain Dealer*, September 11, 1949.

12. "Famous Navy Flier Admits Peddling Gossip," *San Bernardino Sun*, September 7, 1949.

13. Buel and Knebel, "Ohio under the Dome."

14. Barnes, "Adm. Thomas D. Davies."

15. Young, "Persistence Key Word" (see chap. 4, n. 33).

16. "Bailey to Talk of Plane," *San Fernando Valley Times*, March 8, 1969.

17. "Record in '62 Came Easier," *New York Times*, December 22, 1986.

BIBLIOGRAPHY

Archival Sources

AL-7 Col. Clarence "Bill" Irvine Photo Album. San Diego Air and Space Museum, Library & Archives. https://www.flickr.com/photos/sdasmarchives/albums/72157635623864514.

Horner, Edward N. Family papers. Courtesy Kenneth Horner. (ENHP)

National Museum of the United States Air Force. Dayton OH. (NMUSAF)

Naval History and Heritage Command. Washington DC. (NHHC)

Tabeling, Roy H. Family papers. Courtesy LynnAnn Tabeling. (RHTP)

United States Air Force Biographies. https://www.af.mil/About-Us/Biographies/. (USAFB)

U.S. National Archives. College Park MD. (NARA)

U.S. War Department. *Press Releases of the Bureau of Public Relations, United States War Department, 1946.* Washington DC: The Bureau, September 3, 1946–October 31, 1946. Internet Archive, University of Illinois–Urbana-Champaign Collection. https://archive.org/details/immediaterelease1946unit/page/n1.

Published Sources

Aerology Operational Analysis: Aerological Aspects of Operation Turtle, 29 Sept.–1 Oct., 1946. Washington DC: Chief of Naval Operations, Aerology Section, January 1947.

"America by Air: A New Generation of Airliners." Smithsonian National Air and Space Museum. 2007. https://airandspace.si.edu/exhibitions/america-by-air/online/heyday/heyday04.cfm.

"American Aircraft in the RAF." *Flying Magazine.* September 1942.

Anderson, Albert D. "Letters: Dreamboat's Green Light." *Popular Science Monthly* 150, no. 2 (February 1947).

Badrocke, Michael, and Bill Gunston. *Lockheed Aircraft Cutaways: The History of Lockheed Martin.* London: Osprey Aviation, 1998.

Barrett, Richard E. *Aviation in Columbus.* Charleston SC: Arcadia Publishing, 2012.

"Battle of Kansas." *Kansas History* 13, no. 8 (November 1945): 481–85.

"Bausch & Lomb News Release." *Australasian Journal of Optometry* 30, no. 5 (May 1947): 214–17.

Bellamy, John C. "History of Pressure Pattern Navigation." *Navigation* 43, no. 1 (Spring 1996): 1–8.

"Black Cats." *Flying* 35, no. 4 (October 1944): 114–15, 300–306.

Boyle, James M. "This Dreamboat Can Fly!" *Aerospace Historian* 14, no. 2 (Summer 1967): 85–92.

Boyne, Walter J. "The B-29's Battle of Kansas." *Air Force Magazine* 95, no. 2 (February 2012): 94–97.

———. "Von Karman's Way." *Air Force Magazine* 87, no. 1 (January 2004): 74–78.

Broughton, David. "The Aries Flights." In *A History of Navigation in the Royal Air Force*, 73–91. London: Royal Air Force Historical Society, 1997.

Coffey, Thomas M. *Iron Eagle: The Turbulent Life of General Curtis LeMay*. New York: Crown Publishers, 1986.

Cordes, Claire. "Over the Top of the World." *Boys' Life* 37, no. 2 (February 1947).

Crowe, Adm. William J., Jr. With David Chanoff. *The Line of Fire: From Washington to the Gulf, the Politics and Battles of the New Military*. New York: Simon & Schuster, 1993.

Crowson, Delmar L. "Meteorological Aspects of Recent Long-Range Army Flights." *Bulletin of the American Meteorological Society* 28, no. 5 (1947): 220–27.

Curtiz, Michael, dir. *Casablanca*. Burbank: Warner Bros, 1942.

Daso, Dik. *Architects of American Air Supremacy: General Hap Arnold and Dr. Theodore von Kármán*. Maxwell Air Force Base AL: Air University Press, 1997.

———. "Arnold's Evolution: The Legendary Airman Drew Inspiration from Many Sources." *Air Force Magazine* 96, no. 9 (September 2013): 125–29.

———. "Origins of Airpower: Hap Arnold's Early Career in Aviation Technology, 1903–1935." *Airpower Journal* 10, no. 4 (Winter 1996): 70–92.

Davies, Alex. "Thank the Planet's Shifty Magnetic Poles for Runway Renaming." *Wired*, January 19, 2018.

Davies, Cdr. Thomas D., and Lt. Hugh L. Hanson. "In 'Operation Turtle' Design Turned the Trick." *Aviation* 46, no. 1 (January 1947): 70–71.

"The 'Dreamboat' Flight." QST 30, no. 12 (December 1946): 66–68.

Drescher, Cynthia. "Why Airplanes Sometimes Fly over the North Pole." *Condé Nast Traveler*, May 17, 2017.

Duffy, Paul, and Andrei Kandalov. *Tupolev: The Man and His Aircraft*. Warrendale PA: SAE International, 1996.

"Flying Secretary." *Boeing Magazine* 15, no. 12 (December 1945): 3.

Foster, John, Jr. "Science on Wings." *Science Illustrated*, February 1947, 96.

Francillon, René J. *Lockheed Aircraft since 1913*. Annapolis MD: Naval Institute Press, 1987.

Frey, Gary F. "The Hurricane Hunters." *Journal of the American Aviation Historical Society* 25, no. 1 (Spring 1980): 45–56.

Frisbee, John L. *Makers of the United States Air Force*. Washington DC: Air Force History and Museums Program, 1996.

"From War to Peace." In *Of Men and Stars: A History of Lockheed Aircraft Corporation*, chap. 7 (serial, erroneously numbered chapter 8). Burbank CA: Lockheed Aircraft Corporation, October 1957. http://www.mbmcdaniel.com/burbankia/of_men_and_stars_7.pdf.

Gavin, James M. *Airborne Warfare*. Washington DC: Infantry Journal Press, 1947.

Gladwell, Malcolm. *The Bomber Mafia: A Dream, a Temptation, and the Longest Night of the Second World War*. New York: Little, Brown, 2021.

Grant, John. *Encyclopedia of Walt Disney's Animated Characters: From Mickey Mouse to Hercules*. New York: Hyperion, 1998.

Grierson, John. *Challenge to the Poles: Highlights of Arctic and Antarctic Aviation*. London: G. T. Foulis, 1964.

Gulliver, Capt. Victor S. "The Truculent Turtle's Excellent Adventure." *Wings of Gold* 36, no. 1 (Spring 2011): 44–48.

Gurney, Gene. *B-29 Story: The Plane that Won the War*. Greenwich CT: Fawcett, 1963.

Halliday, Hugh A. "The Aries Flights of 1945: Air Force, Part 1." *Legion*, January 1, 2004.

Hamlin, Fred. Report from Washington. *Flying Magazine* 40, no. 6 (June 1947): 34, 89–90.

Harding, Edwin T. "The U.S. Naval Weather Service." In *6th U.S. Navy Symposium on Military Oceanography: The Proceedings of the Symposium*, vol. 1, 17–23. Seattle: Applied Physics Laboratory, University of Washington, 1970.

Hays, Maj. Norman P. "Across the Top of the World." *Air Trails and Science Frontiers* 28, no. 1 (April 1947): 24–28, 92–97.

———. "Report on Navigation of Pacusan *Dreamboat* Flight." *Navigation* 1, no. 6 (June 1947): 142–48.

———. "Wanted: 1000 Navigators." *Flying* 45, no. 3 (September 1949): 40–41.

Hernan, Brian. "The Truculent Turtle." Address of Old Fliers Group, Jandakot, Western Australia, March 2010. https://www.youtube.com/watch?v=_LoxY-hqJXY.

"High Flyer." *Boeing Magazine* 16, no. 6 (June 1946): 6.

"History, 53d Reconnaissance Squadron, Very Long Range, Weather, Grenier Field, Manchester, New Hampshire." Historical Records, Air Weather Service, Air Transport Command, Pt. 1, vol. 2, no. 15 (September 1946). Air Force Historical Research Agency, Maxwell Air Force Base, Alabama.

Horner, Lt. (JG) Edward N., MD. "Takeoff of the 'Truculent Turtle' (Lockheed P2V-1 Neptune), Perth, Western Australia, Sept. 29, 1946 (with Narration from 1990)." https://www.youtube.com/watch?v=8cyQklgmfzE.

Hubbard, Charles J. "The Arctic Isn't So Tough." *Saturday Evening Post*, August 26, 1944.

Huber, Louis R. "B-29 Box Score." *Boeing Magazine* 16, no. 6 (June 1946): 6–7.

———. "Over the Top." *Boeing Magazine* 16, no. 10 (October 1946): 3–5.

Innovation with Purpose: Lockheed Martin's First 100 Years. Bethesda MD: Lockheed Martin Corporation, 2013.

"Inside the Dreamboat." *Popular Science Monthly* 149, no. 6 (December 1946): 91–96.

Irvine, C. S. "The Briefing." *Boeing Magazine* 16, no. 10 (October 1946): 2.

———. "Record Flights Are Made of Mathematics." *Boeing Magazine* 15, no. 12 (December 1945): 3–5, 14.

"It's a Small World." *All Hands*, November 1946.

Jackson, James L. "Aerospace." *American Speech* 41, no. 2 (1966): 158–59.

Jacobsen, Meyers K. "The Red-Tailed Beauties of the 7th Bomb Wing." *American Aviation Historical Society Journal* 24, no. 1 (Spring 1979): 19–40.

Jacobson, Mark Z., Jordan T. Wilkerson, Sathya Balasubramanian, Wayne Cooper, and Nina Mohleji. "The Effects of Rerouting Aircraft around the Arctic Circle on Arctic and Global Climate." *Climatic Change* 115 (2012): 709–24.

"Jitters and Jets." *Of Men and Stars: A History of Lockheed Aircraft Corporation*, chap. 8 (serial). Burbank CA: Lockheed Aircraft Corporation, November 1957. http://www.mbmcdaniel.com/burbankia/of_men_and_stars_8.pdf.

Johansen, Herbert. "What Can Our Bombers Do Now?" *Popular Science Monthly* 153, no. 2 (August 1948): 74–79.

Knott, Richard C. *Black Cat Raiders of World War II*. Annapolis MD: Nautical & Aviation Publishing, 1981.

Leman, Jennifer. "The Magnetic North Pole Is Rapidly Moving Because of Some Blobs." *Popular Mechanics*, May 15, 2020.

LeMay, Gen. Curtis E. *Mission with LeMay: My Story*. With MacKinlay Kantor. Garden City NY: Doubleday, 1965.

LeMay, Curtis E., and Bill Yenne. *Superfortress: The B-29 and American Air Power*. New York: McGraw-Hill, 1988.

Lewis, Isabel M. "The Navigational Stars." *Nature Magazine* 40, no. 4 (April 1947): 211–12, 218.

Livermore, Philip W., Christopher C. Finlay, and Matthew Bayliff. "Recent North Magnetic Pole Acceleration towards Siberia Caused by Flux Lobe Elongation." *Nature Geoscience* 13 (2020): 387–91.

"Lockheed P2V Neptune." *Aero Digest* 66, no. 4 (April 1953): 28–48.

Lucky Bag. Annapolis MD: U.S. Naval Academy, 1937.

Maclure, K. C. "The Artic Flights of the Lancaster 'Aries,' May 1945." *Navigation* 1, no. 1 (March 1946): 3–7.

———. "Polar Navigation." *Arctic* 2, no. 3 (December 1949): 183–94.

———. "Technical Aspects of the Aries Flights." *Geographical Journal* 107, no. 3/4 (March–April 1946): 105–23.

Mahnken, Thomas G. *Technology and the American Way of War Since 1945*. New York: Columbia University Press, 2010.

Markus, Rita M., Nicholas F. Halbeisen, and John F. Fuller. *Air Weather Service: Our Heritage, 1937–1987*. Scott AFB IL: Military Airlift Command, 1987.

Maurer, Maurer, ed. *Air Force Combat Units of World War II*. Washington DC: Office of Air Force History, 1983.

McCullough, David. *Truman*. New York: Touchstone, 1992.

McDowell, Mick. "Howard County's Silver Star Recipients." *Historically Speaking* (Spring 2004): 1–3. http://historichc.org/wp-content/uploads/2017/04/newsletter-vol-008-iss-1.pdf.

McFarland, Stephen L. *A Concise History of the U.S. Air Force*. Washington DC: Air Force History and Museums Program, 1997.

McKinley, D. C. "The Arctic Flights of *Aries*." *Geographical Journal* 107, no. 3/4 (March–April 1946): 90–101.

Meyer, Jeffrey. "Andersen Air Force Base, Guam: 75th Anniversary Issue, USAF Heritage on Guam Pamphlet, 1945–2020." Andersen Air Force Base, Guam: 36th Wing Public Affairs Office, April 2019. https://www.andersen.af.mil/Portals/43/36%20WSA%20TENANT/ComRel/2019-20%20AAFB%2075th%20Pamphlet.pdf.

Moore, Richard. "Interview on Experiences in World War II (Transcript)." World War II Oral Histories Project, Kansas Historical Society, September 6, 2007. https://www.kansasmemory.org/item/218231.

Morison, Samuel Eliot. *The Atlantic Battle Won, May 1943–May 1945*. Vol. 10 of *History of United States Naval Operations in World War II*. Urbana: University of Illinois Press, 2002.

Morrison, Wilbur H. *Birds from Hell: History of the B-29*. Central Point OR: Hellgate Press, 2001.

Myers, Gary P. *The Story of Davis-Monthan AFB, 1940–1976*. Tucson AZ: U.S. Air Force, 1982.

National Park Service. "A Red Bolt from the Blue: Valery Chkalov and the World's First Transpolar Flight." Fort Vancouver National Historic Site. Accessed June 7, 2021. https://www.nps.gov/articles/aredboltfromtheblue.htm.

Neal, Wesley. "Add Miles to Your X-C." *Skyways* 10, no. 2 (February 1951): 26–27, 42–43.

Nelson, Cdr. Christopher. "The Revolt of the Admirals and Today's Battle over the Defense Budget." *U.S. Naval Institute Proceedings* 146, no. 9 (September 2020). https://www.usni.org/magazines/proceedings/2020/september/revolt-admirals-and-todays-battle-over-defense-budget.

O'Connor, Derek. "Flight of the Wellesleys." *Aviation History*, May 2013. https://www.historynet.com/flight-of-the-wellesleys.htm.

Ortensie, R. Ray. "'Battle of Kansas' and the Birth of the Superfortress." *AFMC Flashbacks*, Air Force Materiel Command, August 14, 2018. https://www.afmc.af.mil/News/Article-Display/Article/1602130/.

Patrol Squadron Two Association. "Tales of the 'Truculent' Turtle." Turtle Project, n.d. Accessed June 7, 2021. http://www.patron2.com/files/Turtle/turtleproj.html.

Phillips, Edward H. "Wichita Builds the B-29." *King Air* 10, no. 8 (August 2016): 26–36.

Polmar, Norman. "Historic Aircraft: The First Nuclear Bomber." *Naval History* 17, no. 1 (February 2003): 14–16.

Polmar, Norman, and Christopher P. Cavas. *Navy's Most Wanted: The Top 10 Book of Admirable Admirals, Sleek Submarines, and Other Naval Oddities*. Washington DC: Potomac Books, 2009.

Price, Wesley. "The Air Force Gives Birth to a Miracle." *Saturday Evening Post*, August 25, 1945.

"Professional Notes: Operation Turtle." *U.S. Naval Institute Proceedings* 73, no. 3 (March 1947): 371–73.

Putney, Diane T. "USAF Yearbook—1947." *Air Power History* 44, no. 3 (1997): 4–79.

Rankin, Eugene P. "It Was: One Long Hop: The Flight of the Truculent Turtle." *Naval Aviation News* 79, no. 1 (November–December 1996): 22–23.

———. "Narrowing Horizons." *National Aviation Clinic Proceedings*. Oklahoma City: Times-Journal Publishing, 1946.

Reid, Walter S. "I Rode the Turtle." *Popular Mechanics* 86, no. 6 (December 1946): 89–93, 268, 272, 276.

Rockwell, Norman. "So You Want to See the President!" *Saturday Evening Post*, November 13, 1943.

Rosenberg, David Alan. "American Postwar Air Doctrine and Organization: The Navy Experience." In *Air Power and Warfare: The Proceedings of the 8th Military Symposium, United States Air Force Academy, 18–20 October 1978*, edited by Col. Alfred F. Hurley and Maj. Robert C. Ehrhart, 245–71. Washington DC: Office of Air Force History, Headquarters USAF, and United States Air Force Academy, 1979.

"Sailors Feel Safe When Sighting St. Elmo's Fire." *All Hands*, February 1947.

Salnikov, Yuri. "Handshake across the Arctic Ocean." *Soviet Life* 7, no. 370 (July 1987): 28–30, 47–48.

Schratz, Capt. Paul R. "The Admirals' Revolt." *U.S. Naval Institute Proceedings* 112, no. 2 (February 1986): 64–71.

Sproule, Jean A. "Brief History of the Air Force in Hawaii." Hickam Air Force Base HI: Pacific Air Forces Base Command, n.d. Accessed June 7, 2021. https://aviation.hawaii.gov/airfields-airports/oahu/hickam-fieldair-force-base/brief-history-of-the-air-force-in-hawaii-2/.

Stafford, Edward P. "Flight of the Truculent Turtle." *U.S. Naval Institute Proceedings* 117, no. 8 (August 1991): 45–48.

Stefansson, Vilhjalmur. "The Arctic as an Air Route of the Future." *National Geographic* 42, no. 2 (August 1922): 205–18.

Stevens, Ron. "Truculent Turtle—Sep 1946." *VP-68 Hawk's Nest* 42 (September 2012): 8–11.

Sturma, Michael. *Fremantle's Submarines: How Allied Submariners and Western Australians Helped to Win the War in the Pacific*. Annapolis MD: Naval Institute Press, 2015.

Taylor, Frank J. "Jatos Get 'Em Up." *Saturday Evening Post* 217, no. 47 (May 19, 1945).

Thomas, Lowell, and Lowell Thomas Jr. *Famous First Flights: Sixteen Dramatic Adventures*. New York: Skyhorse Publishing, 2016. E-book.

Thompson, Warren E. "Historic Flight of Betty Jo." *Flight Journal* 11, no. 1 (February 2006): 32–38.

Tillman, Barrett. *LeMay: A Biography*. New York: St. Martin's, 2015. E-book.

Trest, Warren A. "View from the Gallery: Laying to Rest the Admirals' Revolt of 1949." *Air Power History* 42, no. 1 (1995): 16–29.

"'Truculent Turtle' Smashes Long-Distance Record." *U.S. Air Services* 31, no. 10 (October 1946): 18–20.

Truman, Harry S. "Our Armed Forces Must Be Unified." *Collier's Magazine*, August 26, 1944, 16.

U.S. Senate. *Hearings before the Committee on Military Affairs, United States Senate, Seventh-Ninth Congress on S. 84 and S. 1482*. Washington DC: Government Printing Office, 1945.

———. *Study of Airpower: Hearings before the Subcommittee on the Air Force of the Committee on Armed Services, United States Senate, Eighty-Fourth Congress, Second Session, Airpower (Lt. Gen. C. S. Irvine), May 10, 1956, Part 6*. Washington DC: Government Printing Office, 1956.

VP-81 *Album: U.S. Navy Patrol Squadron 81 "Black Cats," Nov. 25, 1943–July 15, 1944*. Anonymous, n.p. http://www.pbycia.org/pbycia.nsf/VP81Album?OpenView&Count=1000.

Whitaker, Wayne. "Meet the B-29!" *Popular Mechanics* 82, no. 2 (August 1944): 8–13.

Whitcomb, Col. Ed. *On Celestial Wings*. Montgomery AL: Air University Press, 1995.

Wilkinson, Stephan. "Sea Sentinel." *Aviation History* 28, no. 6 (July 2018): 36–45.

Wolk, Herman S. *The Struggle for Air Force Independence, 1943–1947*. Washington DC: Air Force History and Museums Program, 1997.

Wylie, Max. *Radio and Television Writing*. New York: Rinehart, 1956.

Yates, Donald N. "We Shall Have Weather." *Army Information Digest* 3, no. 1 (January 1948): 8–16.

INDEX